Praise for *Pressed Plants*

"*A well-dried, well-pressed plant is a thing of beauty that anyone can appreciate. But even more, such a specimen yields vital information about the Earth's vegetation past and present, and can help us understand how to protect plant life—and thus all life—for the future. This tells you everything you need to know to create your own aesthetically pleasing and scientifically informative plant specimens. It is written in a friendly, conversational manner and emphasizes the land stewardship and habitat conservation considerations that inform ethical plant collecting. Not only newcomers to plant collecting but also experienced plant scientists and collections managers should have a copy of this book for their own reference, and to share with others.*"

—**Dr. Barbara M. Thiers**, director emerita and honorary curator, William and Lynda Steere Herbarium of the New York Botanical Garden past president, American Society of Plant Taxonomists and Society for the Preservation of Natural History Collections and research adjunct, Denver Botanic Garden

"*Well-collected specimens are the basis for proper scientific knowledge of plants, but there has recently been a decline in training in this fundamental skill. Linda Lipsen and Derek Tan have now produced a much-needed guide that will serve as a great introduction for botanical newcomers and professionals alike.*"

—**Quentin Cronk**, director, Beaty Biodiversity Museum past director, UBC Botanical Garden and professor, Department of Botany, University of British Columbia

"This is a wonderful and timely guide to making plant specimens and explaining why they matter. Linda's practical and thorough treatment of plant collecting encourages a new generation of botanists and enthusiasts to carry on the tradition of making specimens that further our understanding of the diversity and distribution of the plants around us."

—**David Giblin**, collections manager and research botanist, University of Washington Herbarium, Burke Museum

"This enthusiastic, approachable and helpful book will get you well on your way to a lifetime of appreciating and collecting plants. Packed full of detailed instructions and tips, as well as beautiful and useful illustrations, this guide can be appreciated by both beginners and seasoned plant collectors! The author covers everything you need to know to start your very own scientific plant collection, in cheerful language and in a portable format that you can carry with you on your plant-hunting adventures."

—**Amanda M. Savoie**, research scientist and botany director, Centre for Arctic Knowledge and Exploration, Canadian Museum of Nature

Pressed Plants

Also from the Royal BC Museum

Mushrooms of British Columbia
by Andy MacKinnon and Kem Luther

Wild Flowers
by Emily Carr,
with illustrations by Emily Woods

Nature Guide to the Victoria Region
edited by Ann Nightingale and Claudia Copley

PRESSED PLANTS

Making a Herbarium

Linda P.J. Lipsen
with illustrations by Derek Tan

Victoria, BC

Published by the Royal BC Museum, 675 Belleville Street, Victoria, British Columbia, V8W 9W2, Canada.

The Royal BC Museum is located on the traditional territories of the Lekwungen (Songhees and Xwsepsum Nations). We extend our appreciation for the opportunity to live and learn on this territory.

Cover and interior design by Jeff Werner
Copy editing by Grace Yaginuma
Proofreading by Lynne Melcombe
Index by Stephen Ullstrom

Library and Archives Canada Cataloguing in Publication
Title: Pressed plants : making a herbarium / Linda P.J. Lipsen ; with illustrations by Derek Tan.
Names: Lipsen, Linda P. J., author. | Tan, Derek, illustrator. | Royal British Columbia Museum, issuing body.
Description: Includes bibliographical references and index.
Identifiers: Canadiana (print) 20220286620 | Canadiana (ebook) 20220396566 | ISBN 9780772680563 (softcover) | ISBN 9780772680570 (EPUB)
Subjects: LCSH: Herbaria—Handbooks, manuals, etc. | LCSH: Botanical specimens—Collection and preservation. | LCSH: Botanical specimens—Collection and preservation—Handbooks, manuals, etc. | LCSH: Botanical specimens—Drying. | LCSH: Botanical specimens—Drying—Handbooks, manuals, etc. | LCGFT: Handbooks and manuals.
Classification: LCC QK61 .L57 2023 | DDC 580.74—dc23

10 9 8 7 6 5 4 3 2

Printed and bound in Canada by Friesens.

This book is dedicated to my greatest supporters in life who I dearly love, Michael and Raoul Lipsen

Contents

Acknowledgements

I would like to thank my partner, Michael Lipsen, and my son, Raoul Lipsen, for their support and encouragement. I could not have done this without my friendly reviewers Bev Ramey, Julia Alards-Tomalin and Yukiko Stranger-Galey. I would also like to thank Justin Perry for his invaluable Indigenous perspective in the section about respectful and ethical collecting, and the expert reviews by Ken Marr, Heidi Guest and Eve Rickert.

Introduction

Every day, we experience plants—enjoying them for their smells and tastes and relying on them for our food, clothing, housing and medicines. Plants, and their flowers and fruits, are an integral part of our lives and are crucial for our survival. We are intimately familiar with the plants we find in our own backyards and neighbourhoods, even if we don't know their official scientific names.

So why press plants for science? Because plant biodiversity research needs you! Botanists cannot properly document plant biodiversity on their own, and it takes motivated and curiously minded people like *you* to help build comprehensive research collections. Over the past 50 years, plant collections have seen a steady decline in donations, but in just the last five years we have started to see a revival in people's curiosity to create specimens of the natural world. We all rely on plants, and by collecting and donating plant specimens, we can help scientists to better understand how to protect the world's plant biodiversity.

Collectors and Their Collections

Humans have always been collecting and using plants, yet the idea of collecting a plant, pressing it and attaching it to a piece of paper with its name and details of habitat and location has been around for only about 500 years. It was a way of recording a species' existence along with its variation and location to share with others, as it was not always easy to capture such fine plant details with words or illustrations—and of course a way to capture one's discoveries. Since then, collectors have been documenting plants from all over the world. The passion of plant collectors taking the time to observe nature, collect specimens and donate the diversity they have experienced has led to a wealth of information about the Earth and our intertwined history. We continue to catalogue plant life, which is no small task, since plant distributions and ecosystems change—much to do with human interactions. These early plant collections tell stories of the past, but the present and future story of life on Earth will be built by *you*.

This book is a guide for anyone who is curious about plants and has the desire to preserve them and note their place on Earth. I decided early on to leave algae, lichens, fungi and bryophytes to be covered in their own books, as they are just different enough that they deserve their own spotlight. I will get you on your way to understanding how to collect plants and why, and how to make useful and beautiful specimens. As you gain more experience in collecting, there are other resources available with techniques for pressing specific plant groups and collecting in particular ecosystems, along with a plethora of specialized books, websites and apps for identification.

Look out for the following symbols throughout this book:

Reusable items and sustainable ideas

Tips and tricks

A step-by-step checklist

1 Preparing to Collect

Before you go into the field, or your backyard, you must first get prepared.

You don't need to spend a lot of money to collect and press plants. This activity is inexpensive and easy to enjoy, making this a very inclusive science—get your family, friends, neighbours and community groups involved!

Collecting Tools and Supplies

Here's a list of tools and supplies I use while collecting. Interestingly, the basic tools for collecting plants haven't changed much over the past 100 years, and these are all you need to make a great specimen. I've also included a list of *more* handy tools and supplies that I have found useful in the field, as well as *specialized* tools and supplies if you want to enhance your collection information for research.

Secateurs	**Field notebook**	**Plastic bags**
Hand trowel	**Wooden pencil**	**Plant tags**
Hand lens	**Ruler**	

I picked up my first set of secateurs and trowel at a garage sale. Look out for collecting tools in thrift stores as well.

My favourite alternative tool to a hand trowel is a *hori hori*. I once watched a botanist use its serrated edge to cut through the trunk of a *Yucca* tree (an herbaceous plant that is 10 cm (4") in diameter!) and slice it in half to expose the internal structure. What a strong and diverse tool!

Secateurs (also known as pruners)
A good set of secateurs is essential for cutting plant stems, and branches from trees and shrubs.

Hand trowel
A good sturdy hand trowel will help when digging underground plant material, such as long carrot-like roots, tulip-like bulbs and horizontal underground stems called rhizomes found with many grasses and grass-like plants.

Hand lens (also known as a loupe)
A hand lens is usually 10× magnification or higher. While this is not crucial for collecting, it is the botanist's essential tool for seeing small features of a plant in the field. I string the hand lens around my neck (like a necklace) so I am always prepared to have a quick "look-see" at micro plant features.

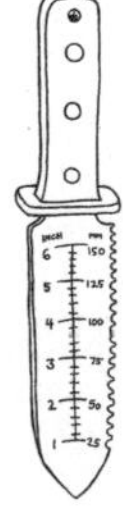

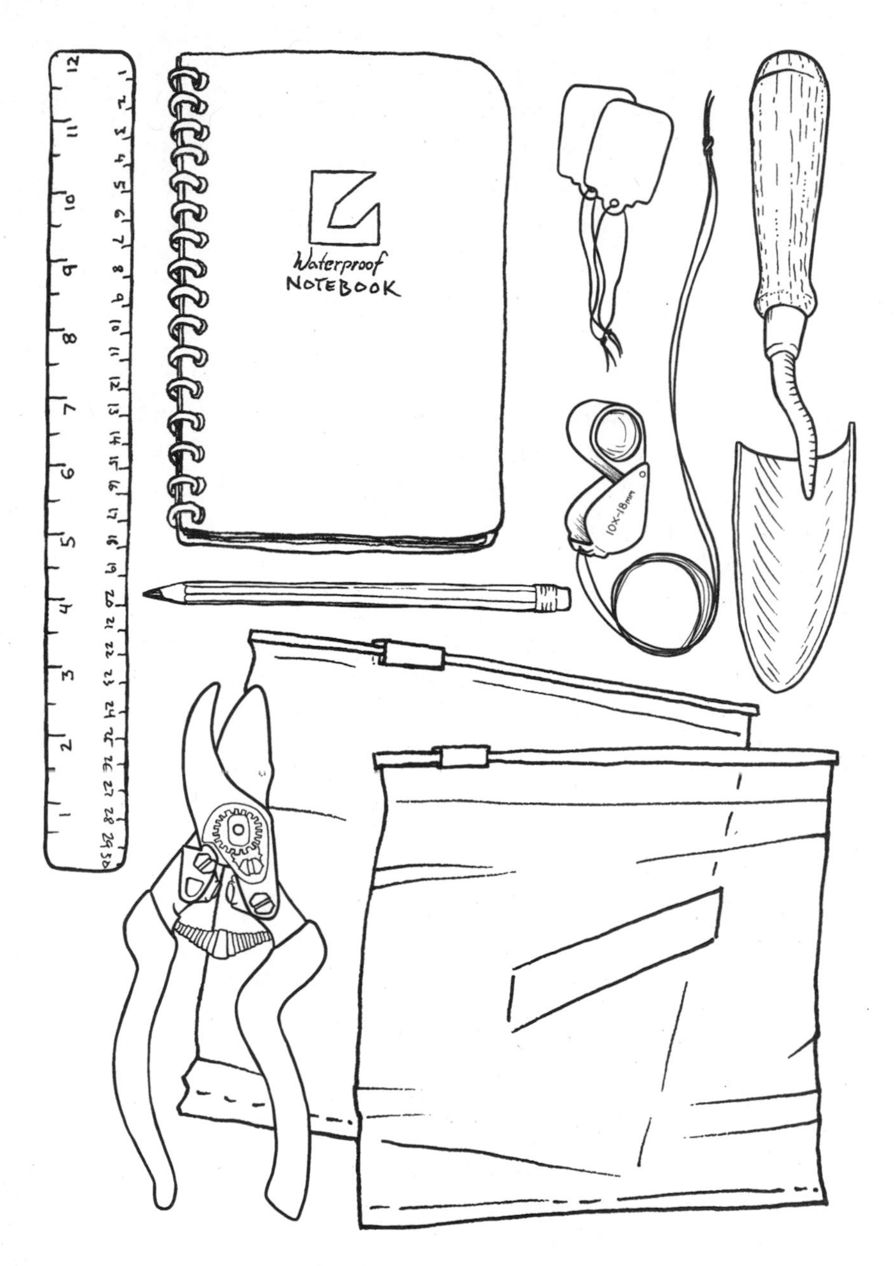

Collecting tools and supplies

Field notebook
A dedicated waterproof field notebook keeps all your collection information in one place. Your field notes are precious, and even if it doesn't rain, your notes will be safe from spills and mould for long-term safekeeping.

Wooden pencil
Why not a pen or mechanical pencil? Great question! Pens often fail because of sand, dirt and moisture, as do mechanical pencils, which are also difficult to load while in the field. Wooden pencils are reliable and easy to sharpen with a hand knife. Ink pens do not write well on waterproof paper.

Ruler
This is crucial in the field to take measurements of plant parts and record them in your field notebook for identification before they dry and shrink or shrivel up. Some Rite in the Rain notebooks have ruler markings on the inside back cover.

You can also use lightweight containers or whatever you happen to have in your pack. In the field I have seen people reuse their empty sandwich, snacks and drink containers to carry even more plant collections back home (after they have eaten lunch, of course)!

Plastic bags
Bring lots of plastic bags! Lugging your big plant press on your back all day is just not a reasonable way to collect. Instead use plastic bags, including zip-lock bags, to temporarily store plant material to transport back to your press.

Plant tags
These are small, stiff pieces of card stock on individual pieces of string. You tie the plant tag to your specimen with your unique collection number that refers back to your field notebook (see "How to Collect," page 35). These are

not a necessity, but over the years I have found these to be useful to track each plant collection when you return from the field.

Make your own plant tag by sliding the plant stem through a pre-made hole in a piece of paper with the unique collection number written on it.

More Handy Tools and Supplies

Small paper bags

These are very handy for bulky items like cones, and you can wrap them around large or dirty roots to transport back for cleaning and pressing. They are also useful for bagging succulents and cacti as they can go mouldy quickly if stored in sealed plastic bags for too long.

Toothbrush

For cleaning mud and dirt off roots of the plant before you press it, as it is difficult to remove after the specimen is dried. Dirt can be problematic as moisture, insects and microbes can hide in it, becoming long-term issues in collections, as well as just plain dirty!

Hand knife

For digging up fine roots and digging plants out of rock beds or tight spaces—and for sharpening your wooden pencil! You can even use a screwdriver, which is more cost effective.

Gloves

Be careful and wear gloves when collecting. They protect your hands from thorns and prickles and also from possible unknown plant threats such as chemical compounds found in poison oak or poison ivy or the sap of some species, which is activated by sunlight and can cause skin blisters. Plants have many ways to deter predators and these same defence mechanisms can cause issues for collectors. Some plants have needle-like hairs or projections like those of *Urtica dioica* (stinging nettles), which can penetrate the skin and cause itchiness, rashes and blistering. Still others, like *Toxicoscordion venenosum* (death camas), are so toxic that if you touch them with your bare hand and then touch your eyes, mouth or nose, you could start to feel quite unwell.

T You may sometimes have to record your location *after* you return from the field. While collecting, there are situations like incoming lightning storms, diminishing light or high tide when I just have to "grab and go." Later when I am safe I'll look up my collecting location using a reliable map resource (an app, online map or paper map).

Paper map

That's right, I suggest you bring a *paper* topographic map. You'll probably use a map app on your phone, but I still advise you to bring a physical map, especially if this is your first trip to the collection site. Many digital maps do not show the level of detail needed for proper collection notes, phones can run out of battery, sometimes satellite signals fail, and electronics do not work well in *all* weather. Bring a paper topographic map. You'll thank me!

Specialized Tools and Supplies

The previous tools and supplies listed will be enough to make a useful and beautiful plant specimen, but there are times when you might want to be more precise about the collection information. You might want to collect specialized groups like cacti or aquatics, or woody species like conifers. You might want to collect more information about the plant's features, such as colours that fade during pressing, or be more precise about the location and habitat, and possibly even collect extra material to extract DNA (deoxyribonucleic acid). If you're planning to make a research-grade specimen, these are some of the extra tools and supplies you might want to bring.

Camera or cellphone

A camera can be very useful for taking pictures of your plant before you press it. Make sure to get shots of the plant's leaves, flowers, colours, shape and any interesting features, as well as the plant's habitat. You can also take a photo of your collection number on the tag or your notebook before you take photos of the plant to be sure you can link the photos with the correct collected specimen.

With a cellphone camera, it's easy to zoom in to identify small features, or you can even put your hand lens in front of the small camera lens to take amazing close-ups!

Many of us have cellphones, which you can use for pictures, notes or even voice recordings of your specimen collection details. Your cellphone photo will also have metadata (GPS coordinates, day, time) embedded within each photo.

Global Navigation Satellite System (GNSS) or a Global Positioning System (GPS)

These are global tracking units that record your precise latitude and longitude. While many phones have some form of location tracking, I would still suggest using an official tracking unit made just for this purpose for accuracy.

Clinometer

A clinometer is used to measure slope. Slope can be an important indicator of particular habitats for some species and is often overlooked when documenting collecting information.

If you turn binoculars around and look through the large lens, they magnify—useful if you forget your hand lens.

Binoculars

Binoculars can help you see way up a tree to note distinguishing features that you are not able to observe on the ground, such as opposite or alternate branching patterns, which can be useful in tree identification.

You could also put the plant material in a coin envelope so that it does not get lost in amongst the silica. These envelopes are also helpful if you want to just use one big bag of silica. You can write the specimen details on the envelope and then put all the envelopes in the same bag.

Small plastic baggies with silica gel desiccant

Silica beads are a drying agent that will help ensure that tissue intended for DNA analysis dries quickly and remains dry. (See "DNA," page 38.) I usually half-fill a 5 × 7.5 cm (2″ × 3″) baggie with silica beads for storing extra material for future DNA extraction.

References

Bring reference books such as field guides, regional floras, identification keys and downloaded local checklists for the region. (See "Identification Resources," page 73.)

Your Safety

Now that you have all your tools and supplies, let's make sure YOU are safe while collecting. Here are a few basics, so you can be sure you have the most fun and much success on your collecting adventures.

* Always make sure you have plenty of water and food, as even a small bottle of water and a snack could save your life.

* Make sure someone knows where you are, even if you are exploring with others. Leave a detailed plan (and stick to it), including your expected return time and/or date, and have a cutoff time when they will need to call search and rescue.
* Have a collecting buddy, as it's safer and much more fun! This is how many early collectors were able to go to remote and dangerous places and bring back such wonderful pressed-plant treasures. There is safety in numbers, and you are less likely to be approached by a bear or other predator. If you sustain a bad injury, a collecting buddy can help by phoning search and rescue.
* Check the local area weather where you plan to collect. Often (especially in high alpine areas) the weather can be quite different from where you are located and will change throughout the day. During my research, I once hiked five miles (8 km) in the desert, which went from about 38°C (100°F) to a high elevation site where it started to snow!
* Check road and trail closures on the way to your collecting spot, and check for summer fire dangers.
* If you plan to collect in more remote and dangerous regions, you can always buy or rent a satellite phone or an inReach for remote communications.
* If you are hiking, always carry rain gear, a basic first aid kit, extra clothing and a space blanket.
* If you plan to go to wilder regions known to have bears, carry bear spray, and read the directions carefully or you could end up spraying yourself!

Responsible and Ethical Collecting

Do your due diligence: understand protective agreements and apply for permits before you collect. While these actions did not always occur in the past, now is the time collectors need to *respect where you collect.*

Permits and Agreements

If you are collecting in your neighbourhood, always make sure to ask before you collect on someone's private property. I sometimes make a thank-you gift, like a card with flowers pressed from their garden.

If you plan to collect in a designated parkland, you must contact the park authority well in advance of collecting to request a permit. This not only helps the land managers to better protect habitats and species, but also allows them to report on the variety of activities that take place on these lands. Often they require that you reciprocate your permit acquisition by submitting a list of the species you collected or encountered on your trip. Acquiring a permit to collect may take several months, so submit your request well in advance of your desired collection time.

You should also search out local resources to better understand the rare and endangered species of your collecting area. Check out government resources, local native plant societies or nature groups and local or regional university and college herbariums. Before I collect locally in my region, I always refer to the Conservation Data Centre (CDC) of British Columbia. I look up potentially threatened or endangered species and their ecosystems that I might encounter while collecting to ensure I do not unintentionally collect a rare species. In fact, it's illegal to collect species at risk, and the fact that they are rare should deter you from adding to their demise by collecting them. If I do run across a rare species, I can help the CDC by documenting the site with an observation (see "Incidental Observations," page 33).

If you travel far and intend to collect and bring plants back home, remember there are global agreements in place to protect species, including their cultural significance, economic value and genetic material. The Convention on International Trade in Endangered Species of Wild Fauna and Flora (CITES) is an international agreement in effect since 1975. I use this resource often in my work when researchers want me to send them a specimen and I first must confirm if the species is protected under this act. If you plan on collecting any species in the orchid or cactus plant families anywhere in the world, for example, you will need to apply for a CITES permit, which involves submitting various collection permits, complying with reporting requirements and understanding that bringing any biological collection from another country without the proper permits is illegal. (There is always a threat of bringing a new pest or pathogen that may be hiding between the leaves or in any attached soil.)

Respectful and Ethical Collecting

Indigenous Peoples have been the caretakers on Turtle Island (North America) since time immemorial, and they have developed specific practices in relation to the sustainable harvest of plants. It would be wise to keep these practices in mind as you collect plants, to maintain balance with the ecosystems and ensure your impacts are minimized. Respecting where you collect includes respecting the Indigenous Peoples of these lands and their ways of knowing and doing.

Before collecting on any Indigenous Lands, which is all lands, you should work to build a relationship with local Indigenous communities. Personally, I wouldn't want to collect on Indigenous Land without creating a relationship first, as this is an important part of building trust and respect. However, this may not always be practical. Local band offices may simply not have time to manage every request. If you are collecting plants for recreation or study, the above step is not mandatory.

As a rule of thumb, I suggest reaching out to local Indigenous communities in the following situations:

* When collecting plants on culturally significant sites (e.g., reserve land and areas used for ceremony)
* When collecting plants for commercial use, or when you will be regularly collecting specimens from the same location

Also, there may be areas where it is simply inappropriate to collect (such as land that is sacred or used to grow plants for ceremonial purposes). These should be respected and left untouched.

If you have received permission to collect from the Nation(s), give back and give thanks and acknowledge them in your collection data. These steps show respect to the peoples and the lands you collect from.

There are over 630 Indigenous communities in what is now known as Canada, and each have different yet sometimes similar ways of viewing life and plant collection. It is best not to assume anything about a community's culture or plant collection methodologies. Each community has its own unique culture that we can humbly seek to learn and benefit from.

Plants are not things but living beings that deserve proper respect and care. We share the same wheel of life, and when people harvest plants, we should be mindful that we are impacting another living entity. To offset our impacts, we should aim for a relationship of reciprocity with plants—for example, planting a native species in exchange for the one you took.

To understand this process, we can look to guidance from the practice of the Honorable Harvest, as synthesized by Dr. Robin Wall Kimmerer in *Braiding Sweetgrass* (2013, p.183), a book that builds on many decades of work by Indigenous scholars:

Know the ways of the ones who take care of you, so that you may take care of them.

Introduce yourself. Be accountable as the one who comes asking for life.

Ask permission before taking. Abide by the answer.

Never take the first. Never take the last.

Take only what you need.

Take only that which is given.

Never take more than half. Leave some for others.

Harvest in a way that minimizes harm.

Use it respectfully. Never waste what you have taken.

Share.

Give thanks for what you have been given.

Give a gift, in reciprocity for what you have taken.

Sustain the ones who sustain you and the Earth will last forever.

When collecting, we all must be aware of and sensitive to Indigenous Lands, perspectives and practices. There is a variety of plant uses in many Indigenous cultures around the world including food, medicine, rituals and ceremonies. Indigenous knowledge has been undervalued and unrecognized for too long. There are lots of examples of stolen Indigenous knowledge related to plants that still isn't fully credited to this day. Over hundreds of years, many plant species, along with the knowledge of growing, harvesting and using them, were stolen from Indigenous Peoples and used for the prosperity of the thieves.

2 Collecting Plants

Why, What, When and How?

One of the first questions people ask me is "How do I collect?" Which is not just about how to collect but also why we collect, what to collect and when to collect. I appreciate this question as it means we are now at the really fun stage—collecting itself!

Why Collect?

People collect for many reasons: for crafts, local knowledge, teaching, tracking biodiversity or just the experience and pure joy of pressing plants and recording history! My first collection dates back to my passion for pressing flowers for decoration, crafts and gifts—there is nothing lovelier than a handmade card with pressed flowers. Maybe you want to get to know your "backyard biodiversity" as there is something so satisfying about walking around and learning the names of the plants, maybe even a few insects, birds or animals and other natural wonders, that surround our everyday lives. It's what makes this world familiar and comforting, and I take great pride in sharing the little details and stories about my neighbourhood with others.

Another important reason to collect is the power of a good specimen. I can't tell you how many times plant specimens have been used for documenting a new population or species, for conservation or even for a new understanding of history—of the species, local people or the collector—long forgotten. A specimen can open a window into the history of the collectors who have come before us and documented the plants they learned about throughout their lives.

When I pressed my first plants, I was never quite happy with the results. The specimens turned black or they moulded, they were missing important features for identification, or I had pressed them so poorly they were useless. I wondered, How can this flattened piece of dead plant matter represent the beautiful plant I collected in the field? Why bother? Why not just take a picture and move on? But a specimen is much more useful than a picture. Though the plant is not in its full glory as a specimen, specimens can be scientifically useful and look beautiful if collected and pressed correctly. You can study individual plant features under a microscope, you can remove and rehydrate parts of the plant to re-examine them as if they were alive, and you can even extract DNA. How cool is that! I remind collectors that these specimens are not a full representation of the plant in the field but rather a snapshot of the plant, and its parts in particular, and if collected and pressed properly, this specimen will be useful to many people beyond your lifetime.

When you are collecting, think far into the future—20 years . . . 50, 100 and now 500 years (can you do it?). What story are you telling about this plant: what did it look like, what did it smell like, where was it growing, what other species were surrounding it? As the creator of a specimen, you are the *storyteller* of this plant, painting a picture of its place in the world at this moment in time and space, leaving behind its history for the world to discover and enjoy.

What to Collect

THE THREE "WHAT'S" WHEN YOU COLLECT:

What **species** to collect?	What **plant parts** to collect?	What **information** to collect?

What Species to Collect

What you collect depends on your goal. If you want to pick flowers and press plant parts for crafts or a small personal collection, then you have the freedom to responsibly collect whatever you may find. Most collectors and researchers have a goal or question in mind when collecting, and this will often guide their collection habits.

Over the years I have found that many collectors start off with a theme, which will flourish as their knowledge and comfort with plant identification, habitats and ecosystems grow. Here are a few ideas of where you could start.

Local neighbourhood and backyards

Many collectors start off in their local neighbourhood or backyards since these are accessible and familiar. Even though a new collector might think, "Is that really important?" I would respond, "It's incredibly important. If you don't, who will?" You will be the *storyteller* who records the changes in your local landscape and documents your backyard's biodiversity. These types of collections, with consistent samples from a local and reliable collector, can be very informative in detecting changes over time.

First in bloom, last in bloom

A fun collection to make is "first in bloom, last in bloom." Who doesn't love first in bloom, as it signifies spring is on its way, while autumn's last in bloom is a sign that winter is just around the corner? Many botanical gardens and herbariums have been documenting "first in bloom, last in bloom" for decades, and they now find this information useful in the study of global climate change.

Invasive, exotic or introduced
I encourage collectors to document invasive, exotic and introduced species. These types of collections are essential for understanding when certain species arrive, how they disperse and how fast they can change an ecosystem. Surprisingly, some of our new species introductions have come from garden plants escaping their local area. They can often dominate and outcompete native species in the wild quite rapidly.

Range extensions
Once they start to get a feel for species and their range distribution, many collectors discover new range extensions. This is when you find a new population that has never been documented before outside the species' known range. Range extensions can inform us about possible new species habitats that are due to climate change, among other reasons, and highlight new environments where we might expect to find a species. This information can help when working on conservation strategies, land management and restoration projects.

Hybrids and hybrid swarms
As your plant identification skills improve, or if you just have a sharp sense for patterns, shapes and colours, you might want to hunt for hybrids. Hybrids occur when two different species produce offspring. Sometimes these offspring are *sterile*, meaning they will not produce viable seeds, but sometimes they do produce viable seeds and a new hybrid species can arise.

While we have known and documented naturally occurring hybrids over the centuries, present-day research puts more focused attention on finding and documenting them as they sometimes result in *hybrid swarms*. This is when a hybrid becomes a rigorous and dominating species. Hybrid swarms are important to find and document early, as given enough time (seasons), the hybrid can reproduce and disperse seeds, possibly outcompeting and displacing the parent species and potentially leading to extinction.

Conifers, trees and shrubs

Many collectors don't find these groups quite as esthetically pleasing to collect or as easy to press as herbaceous perennials and annuals, but these are important groups to collect. Conifers, trees and shrubs are the backbone of many ecosystems, providing canopy and shade for a variety of understorey plants. If no one is collecting these specimens, then we are missing important information about the long-term stability of many local ecosystems.

Aquatic plants

Plants in water are often ignored, as they are difficult to collect and press, hard to identify and just plain wet! Plants growing in these habitats are easily overlooked and thus are under-represented in collections. Watch for plants growing in *ephemeral* (short-lived) habitats, like ponds that dry up in the summer, as these have a short blooming time and are often missed when collecting. Also, many new or exotic plant species arrive and move through our waterways and invade without anyone noticing. So, it's time to get your feet wet and start documenting!

What Plant Parts to Collect

The short answer is reproductive parts with intact leaves on the stems. Now of course, there is more to collecting than that. For centuries, botanists organized plant groups based mainly on their reproductive structures, such as flowers and fruits. This was the best way to understand and catalogue historical and hierarchical relationships between plants. Now we can include much more information, including DNA, which greatly aids in understanding the classification of plants.

In general, for *herbaceous* (non-woody) and aquatic plants, you should try to collect the whole plant, from the bottom of the roots to the tippy-top part of the plant. You will include the roots (see "Roots or underground stems" below), all the different leaf types, the whole inflorescence (which includes the main stem, flower buds and mature flowers, and leaf bracts) and sometimes fruit, if available.

For conifers, woody trees and shrubs, cut a portion of the plant the length of an herbarium sheet that represents all the important plant parts needed for identification, including flowers, fruits or cones. This would be 40.5 cm (16″), depending on the newsprint you use.

Remember, if you are going to collect it, collect it right. It's not worth collecting a plant and essentially killing it for a poor specimen that cannot be used because you had too little material or you didn't collect the essential parts for identification.

To learn exactly what parts of a plant are needed, look through local plant guides, online resources and regional taxonomic floras to see what plant parts are important to identify the plant family or group you are collecting.

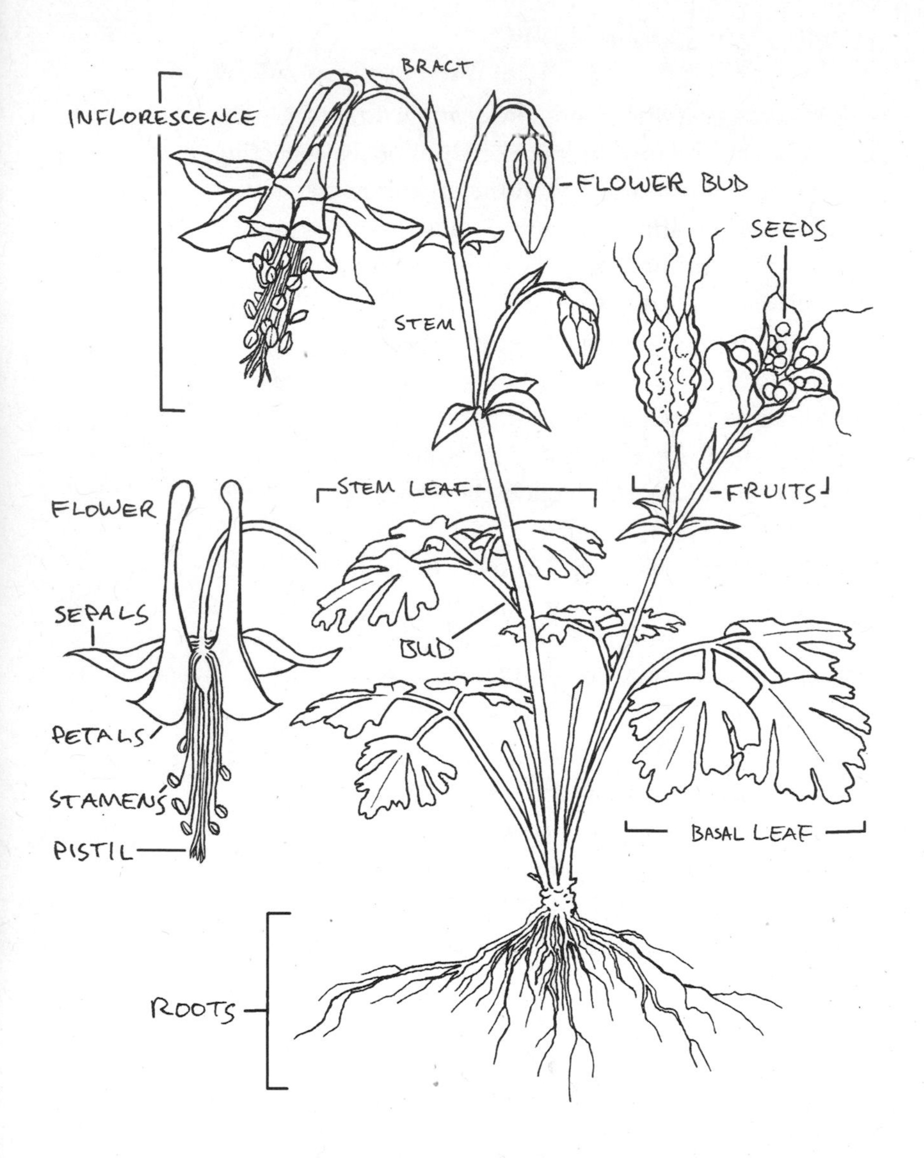

***Aquilegia formosa* in the field with labelled plant parts**

Roots or underground stems
These can be very important as a diagnostic tool to understand if a plant is an annual, biennial or perennial. They are also very useful in the identification of grasses and sedges. It is particularly important to include *rhizomes* (horizontal underground stems) in the collection if they are present. If you know you do not need the roots to identify your species, make sure to cut the specimen in a way to leave enough plant behind to regrow.

Buds
Ah, buds—my favourite part of a plant, as they are full of potential! They can become whatever the plant needs at their time of development (a stem, leaf, flower or fruit) and therefore are the most versatile of all the plant parts. Collectors often forget buds, particularly for woody plants, yet they offer diagnostic features, such as their number, arrangement and position on a plant.

Stems
Even though all the important plant parts are attached to the stem, so you will have already collected it, remember that a stem contains a lot of information, too. This can include its colours, textures or smells, as well as the size of its *lenticels* (pores) used for gas exchange.

Leaves
Make sure to collect all the different types of leaves found on a whole plant. This includes *basal* (base of plant) and stem (along the stem) leaves, as well as new, young leaves near the tippy-top part of a plant. Each can have very different shapes and sizes with varying hair types, depending on where they

grow on a plant and their stage of development (young or old).

Collect enough flowers so you can open up a few to see all the reproductive features for identification in the field, as well as extras for later use.

Inflorescence and flowers

The flowers should be attached to the main inflorescence stem, with all the reproductive parts developed, including *sepals* (often green leaflike parts found on the outside of the petals), petals, and *stamens* and *pistil(s)* (male and female reproductive parts, respectively).

Try to collect fruits and seeds from the same individual you collected flowers from. If that is not possible, collect flowers and fruits from individuals nearby (same population) that are in different stages of maturity, but give this a new collection number, and be clear in your notebook (and label) that these are different individuals and should not be mounted together. If all the plants nearby are at the same stage and lack mature fruits, you can also return to the site later in the season to re-collect.

Fruits and seeds

These are both very important identification features, and yet they are often forgotten when collecting as we are all distracted by the showy flowers! Fruits (ovaries) and seeds (ovules) are part of a plant's reproductive system. Fruits are defined by whether they enclose the seed(s) of a plant, which are the next generation of that species. Their shape, size, design, colour and the way they grow tell us a wealth of information about a plant's evolutionary history and ancestry.

Both parts can also inform us about the plant's method of fertilization and dispersal. Cute little fruit "boats" found in the pea family tell us their seeds are dispersed by water, while barbs on the outside of other seeds can inform us that they're dispersed on the fur of animals, which also inspired the invention of Velcro!

And remember, not all plants have seeds, and not all plants with seeds have fruits. For example, ferns disperse with spores, and conifers have "naked" seeds in their cones (not enclosed in a fruit).

What Information to Collect

The information you collect is the story of this specimen, and you, the storyteller. Think of yourself as a time traveller—when you pick and collect this specimen, you are stopping time. It's a lot of work to collect, and there are plants everywhere, so what made you say, "Yup, that's the one!"? The information you collect will go on the specimen label, which is adhered to the specimen's mounting sheet and tells its story. Tell us about this plant, its surroundings and the other species you see overhead and nearby. Paint a picture, tell a story. Tell us why you collected this individual so others can learn more about this plant, and about you as a collector.

Here is a list of typical collection information found on a specimen label:

Family
Genus and species
Location
Geolocation (latitude & longitude)
Elevation
Habitat
Date collected
Notes
Name of the collector(s)
Collection number
Identified by
Identification date

Many people think if you collect a plant, then you must know what the species is, and that is sometimes not the case. It's OK to have a specimen that you cannot identify easily, as often the ones we struggle with are the ones that we later realize are a new variety, subspecies or even a species new to science! Try your best to name the species, as that is an important part of a specimen's value, but add a "?" if you are not confident in your identification. It's always better to be clear that you are not sure (and maybe write a note about why) than to put an incorrect name to a specimen.

Family, genus and species

The combination of genus and species (or what botanists call a *specific epithet*) along with the plant name author (e.g., "Fisch." for Friedrich Ernst Ludwig von Fischer, who wrote and published the first description of *Aquilegia formosa*) is a plant's full scientific name. In the vascular plant herbarium, we often identify, catalogue and organize at the family level, so you should also add the family to the plant label.

Location

Describe where the collecting site is and what route you took to get there. This description will often include the country, province or state, Indigenous territory, sometimes county or regional boundaries, and a general description of your route. Try to think long term, and avoid using place markers that can change, like a restaurant parking lot or corner gas station, as neither will probably exist in 100 years! Try to use a name that is recognized in an official regional, state or provincial register of geographical names.

Geolocation

Record the coordinates of your collection site in latitude and longitude, or in UTM (Universal Transverse Mercator). These are a more precise location of where you are on Earth, since lots of markers, including geological ones, can change over time. If you are using a GPS or GNSS, remember to record what data you are using (the frame of reference for mapping—WGS 84, NAD 27 or NAD 83), and also record the accuracy of your coordinates in your notebook.

Elevation

Record the elevation for your specimen in either metres (m) or feet (ft), and be consistent. Elevation is an important part of your location description, as high elevation collections are particularly valuable in many plant studies.

Habitat

Describe the habitat surrounding the plant, and any prominent or interesting species growing nearby. This is also where you can add extra information—geological features such as soil or rock type, geographical features such as waterways, or the slope's steepness or aspect (the direction it is facing).

Date collected

Record the day, month and year you collected this plant. Write out the month in full, or at least its first three letters (e.g., Aug = August), and write all four digits of the year. The date is critical to a scientifically useful specimen.

I have seen much confusion with the year, month, and day being switched, and even the wrong century if the year was not made clear by the collector. My favourite is when the collection year could not have been possible unless the collector was still collecting after they had died!

Notes

I encourage you as the storyteller to use this section generously. This is where you are going to tell why you collected this plant so that others treat it with the respect it deserves. There are many different reasons to collect a plant, and the notes section is the most interesting (and often ignored) part of collection information. You can describe any distinguishing features, such as the height of the plant, flower colour and smell, per cent blooming and per cent fruiting. Does the plant have a late or early blooming time or a colour variant? Were any pollinators or seed dispersers visiting? Does the specimen represent a range extension? If so, estimate the number of individuals found in the population, especially if the species is rare, as this is crucial information for conservation. Maybe add something about features that can't be seen with the eyes (such as scent) and those that will potentially change over time (such as flower colour).

Name of the collector(s)
Record who collected the plant. Give yourself credit for all your work, and include any other collecting buddies who took part in your adventures. The collector's name is important for a scientifically useful specimen. Make sure to clearly write out all names, as many people have similar names, and this is for history!

Collection number
Record your unique collection number. This is *your* unique number as a collector, and it correlates to *this* specific specimen you have collected. My suggestion is to have sequential numbers as you go forward through your collecting career, as this will give you the total number you have collected in your lifetime. You can also start off each year at "1" combined with the date (e.g., 20210607.01) or some version of this that works for your collecting habits. This number will be crucial for tracking and understanding the history of your collections—and when I say history, some collectors' collection numbers have reached the hundreds of thousands, representing 40, 50, 60 years or more of collecting!

PLANTS OF BRITISH COLUMBIA, CANADA
LINDA LIPSEN'S HERBARIUM

FAMILY: RANUNCULACEAE
GENUS SPECIES: *Aquilegia formosa* Fisch.

LOCATION: Vancouver Island, Capital Regional District, Fairy Lake Forest Recreation Site, Pacheedaht Territory
GEOLOCATION: LATITUDE: 48.587° N LONGITUDE: -124.350° W
ELEVATION: 16 metres
HABITAT: On a rocky bank, west side of the river
DATE COLLECTED: 2021, June 07
NOTES: 100-150 plants in population, 50 blooming plants red and yellow petals
NAME OF THE COLLECTOR(S): Linda P.J. Lipsen
COLLECTION NUMBER: 413
IDENTIFIED BY: Erin Manton
IDENTIFICATION DATE: 2021, December

A specimen label with collection information

Identified by
Record the person who identified this individual plant—most often yourself but not always. Sometimes you might collect a specimen that is not easily identified and will need to consult with an expert.

Identification date
Record the date the specimen was identified or identification was confirmed. This can be after the collection date, as we do our best to identify the plants in the field but that can't always be the case. Often expert botanists will return to the herbarium with their specimens and use its vast collections to compare identified specimens against their own pressed plants before confirming the field identification.

When to Collect

At high elevations you can often find a species at different stages of blooming by looking for it where the snow has just melted (for early stage), near water runoff, on the shady side of large rocks or on fully exposed, south-facing slopes (for late stage).

Almost any plant collecting manual will tell you to collect when it is dry outside. This is great in theory, but I am from the Pacific Northwest, and that would mean I would only be able to press three months out of the year, if I am lucky. Collecting when it is dry outside will result in a better pressing, but don't let the weather limit you. For best results, try to collect midday, as there will be less morning dew and more fully open flowers. (Many close at night to retain moisture, trap pollinators or protect flowers from predators.) You might have to collect in non-ideal conditions—that's life, as we don't control the weather!

You ideally want to collect when plants are in full bloom, and hopefully with some fully developed fruits, especially if the fruits are needed for identification, as they often are. Also, consider collecting the same plant species during different seasons in the same year to capture the two important reproductive structures (flowers and fruits) along with changes in size, shape and colour of leaves (particularly fall colour). A species can be in full bloom at the coast, while once you go up in elevation you will find blooming to be delayed—and with shorter flowering times, you will need to time your visits precisely. Remember, high alpine areas are sensitive and short lived for blooming.

When Not to Collect

In the plant-collecting world, we have what we call the 1 in 20 rule. If you don't see at least 20 blooming individuals, then you should not collect. This is roughly the number of individuals needed to maintain the genetic diversity of a population. Remember, these plants not only are interesting to us but also provide essential food and shelter. Animals, insects, fungi and a whole host of other organisms rely on these plants for survival.

There are particular groups and families I would discourage you from collecting. These groups include rare and endangered plants and plant families such as lily-like flowers, irises and orchids. While they can be abundant, they often grow in sensitive habitats and should be avoided to err on the side of caution. These are some of our most beautiful flowers, but they are also quite sensitive to disturbance and struggle in most environments to even exist. It is not to say that you will never be allowed to collect these lovely species, but do it with permission, thoughtfulness and care.

Incidental Observations

Many times, when surveying rare and endangered species, you will not be allowed to collect a specimen. Instead, you will take all the data you can and report your observation of the population and site to the BC Conservation Data Centre (www2.gov.bc.ca/gov/content/environment/plants-animals-ecosystems/conservation-data-centre/submit-data). This just means you collect information for an observational report instead of collecting the specimen. Take some diagnostic photos of the plant's important features, document features of the population and habitat and note the coordinates in an Incidental Observation report. This way you can help document the species without having a negative impact.

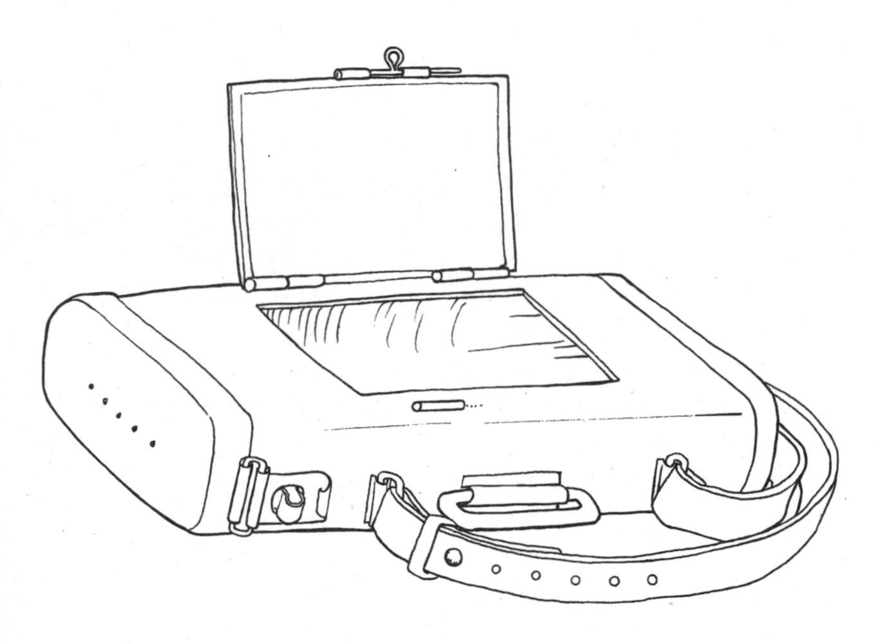

A vasculum

Even in the early days, many collectors used some sort of vessel to carry their collections while they hunted for plants instead of carrying a press all day. Once they got back to home base or camp, they would bring out the press and begin the job of pressing their plants from the day. Eventually this vessel was made specifically for plant collecting and is known as a *vasculum*. While it used to be a smart way to carry collected plants without crushing them, it's not secure and has limited space. You can collect and carry a lot more plants with just a small backpack and lots of baggies!

How to Collect

To me this is the easiest and most enjoyable part of collecting. You get to walk around looking for cool plants to collect, and every time it's different. When I go out, I gather my bag with its tools and supplies and head out until I reach a location where I see a few interesting plants I want to collect.

- ✔ First, I get out my notebook and write at the top of a new page the date, location, geolocation, elevation and habitat of the collection area, as this will now be the data for different plants I intend to collect.

- ✔ I then have a good look at each plant I'm collecting. I note the flower colour and smell, and use my hand lens. I record the number of sepals, petals, stamens and pistils, any distinguishing features about the fruits, and any leaf or stem-hair features. This is also where I take any diagnostic measurements needed for identification.

- ✔ If I have a camera with me, I take a few photos of the plant to show the important features before collecting.

- ✔ Now I assign a unique collection number for this particular plant and tie the plant tag with the collection number around the plant's stem. (Note that plants from the same population will have the same collection number.)

- ✔ Next, I record the collection number in my notebook and write down a preliminary species identification. I also add notes about how common the plant is in the area, and any other descriptions specific to the plant and its habitat.

- Then I collect each plant by either digging up the roots (if needed for identification) or cutting the plant stem near the base, making sure to include basal leaves but leaving enough to allow the plant to regrow. For trees and shrubs, I cut off the end of a branch.

- If I want extra leaves for future DNA extraction, I take a few of them off the plant, place them into a little plastic baggie filled with silica beads, and write the collection number on the outside.

- I place each species individually into a closed plastic bag to temporarily store the plant material. If I am collecting larger plant parts such as cones and larger fruits, or cacti or other plants that are more succulent, I place them into a small *paper* bag to reduce humidity issues that occur with plastic. Often these types of plant parts will need longer to process and dry and are less likely to mould in a paper bag. You can also write the collection number on the outside of the bag to ensure accuracy.

- I make sure to not mix different species or different populations of the same species within a bag, as this can lead to confusion of plant parts that fall off in the bag. Mixing species at any stage—collecting, drying, mounting or cataloguing—can make things difficult and confusing for future studies.

T Before I close up the plastic bag with the collected plant inside, I usually blow into the bag to inflate it a bit and sometimes put a moist paper towel inside if I know I will be out in the field all day. This will help to keep the cells hydrated while the plant is waiting to be pressed. If you run out of room within your collecting backpack, you can also hang the bags on the outside so you can collect more.

T For long day trips or overnight trips, you would have to bring along your plant press as the plants will begin to wilt by the end of your day (or sooner). You could streamline it and reduce its weight by carrying only the cardboard dividers plus newsprint, with string to cinch it all together. The cardboard will likely fit inside a backpack immediately against the back support so it keeps its shape.

I continue collecting additional plants of interest. Then I pack up, move on, find my next collecting site and repeat.

Extra Steps You Can Take

Illustrations

When you are in the field and taking notes about small plant features such as the reproductive parts found inside the flowers, you can also make a rough sketch or illustration in your notebook. Drawing can allow your eye and hand to communicate distinguishing and unique features more readily than a photo. Later in the day when you get to pressing the plants, these important parts might be closed or wilted or even have fallen off, leaving less material to press and look at when trying to confirm your preliminary identification. If you have properly captured the diagnostic features of your plant while in the field, you won't end up possibly destroying any of your beautiful pressed plant when confirming the identification.

Diagnostic photographs

It is always a good idea to take photos when collecting. Often the shape, colour and likeness of the plant is lost in the pressing process, but a picture can help to properly document what you saw. Sometimes a photo or two showing the habitat can help if you are in a hurry, but make sure they are good enough to rely on for writing descriptions. You can also share your photos with community science sites on biodiversity, such as iNaturalist. This allows you to share the specimen and get input from the wider community on your identification and collection. Take a few photos of the plant's overall form, and then close-ups of flowers, seed heads, the stem and leaves (including basal leaves)—all the important features that you want to be part of your specimen.

DNA

If you have pressed, dried and stored your specimen properly, then you have already preserved the DNA for researchers to extract. But if you are thinking for the very long term, take a few extra sets of leaves and place them in an archival plastic bag with silica beads (a drying agent), which will help ensure extended preservation of the valuable DNA. The added benefit of doing this is that scientists will not have to remove parts of your plant for DNA or chemical studies, leaving your specimen intact for longer. If the baggie is small enough, you can place it into a fragment packet (see "How to Mount Your Pressed Plants," page 55) and store it with the specimen, or you can store it in a separate place altogether, like in a shoebox. In herbariums, specimen DNA is often stored in a specific cabinet or freezer.

3 Pressing and Drying Your Plants

Now that you have collected all your plants in their individual bags with their vital identification features (beautiful flowers, delicate leaves and squishy fruits) and the all-important field notes, the next step is to press and dry the plants to create a specimen. To me, this is often the hard part of collecting. It takes time, patience and a certain feel and touch to make a useful yet artistic specimen. Honestly, it will be trial and error, as each botanical family or species has special considerations during this drying process. Sometimes an individual plant just won't behave no matter how much you try to work with it! Once you learn these techniques and have practice with lots of different plants, you too will learn to make beautiful and scientifically useful specimens.

Pressing Supplies

You can also make your own specialized press depending on your interests and needs, such as a smaller press for hiking or small plants.

First, of course, you will need a plant press with all the important bits!

A standard-size herbarium press and its components is 30.5 × 45.7 cm (12″ × 18″) and will include:

2 lattice press backings
2 straps with locking closures
Corrugated cardboard
Newsprint
Blotter paper

Lattice press backings
A plant press should be portable, long lasting and able to withstand pressure. I often use presses made from western red cedar as it is local and withstands rotting, deters bugs and is lightweight and flexible (and therefore won't break from pressure). I have also seen presses made from oak or other woods that will work just as well.

Start where most of us did and simply use large books from around your house! You can also use plywood, although it doesn't ventilate as well as lattice. Check thrift stores and yard sales for old presses because if made well, they can last hundreds of years.

Straps with locking closures
There are many types, and it seems every collector chooses a system they like best. There are nylon or cotton straps in differing widths, with plastic, metal-cam or metal spring-loaded buckles. Remember, the length of your pressing straps limits the height of your press and how much it can expand when filled with plants. I would suggest between 2–3 m (6.5′–10′) in length as I have filled my press so tall it was up to my hip—about 90 cm (3′) high.

These next three items are crucial as they will determine how well each of your plants will press and dry. Also, these pressing items will limit how many plants you can collect, so you should always bring more than you think. When the plants are in bloom, it will be hard not to collect more plants, and more and even more!

Instead of these straps, you can use rope or twine you find around the house to press the boards together. Whichever you choose, make sure that you can tighten it down to ensure even pressure on the backing boards across the press.

Corrugated cardboard

What we in the plant-pressing world call corrugated ventilators is one of your most essential items. The pieces of corrugated cardboard do a few essential functions in the press. First, they separate your specimens and keep them organized; second, they ensure a gentle but even pressure; and third, they allow air to flow through the cardboard for even drying across the specimens.

It is critical for the cardboard to ventilate properly. For this to occur, you must have the long 45.7 cm (18″) edge be the ventilating side and be able to see the Cs of the cardboard. Think of air being able to easily flow through the cardboard along these channels.

I never thought I would care so much about cardboard, but over the years I have become a real cardboard-pressing snob, and these are my tips:

* Don't use old cardboard where the Cs are collapsed or the edges are closed off to air flow.
* Don't use cardboard that has lost its strength and is flimsy, bent or soft.

This is what keeps your plants flat while drying, so don't skimp here!

Newsprint

Try to find 29 × 42 cm (11½″ × 16½″) sheets as they will fit in the press best. Since newspaper comes folded (and is therefore twice the size), it helps to contain specimen movement so you don't lose plant material when it comes time to process your collection. You can either buy blank newsprint paper or use

Make your own corrugated ventilators out of cardboard boxes from your recycling.

printed newspaper (cut to the correct size if necessary); try to use newspaper with no colour pictures—they tend not to absorb moisture as well. Always bring clean newsprint you haven't previously used for pressing to press each specimen. As much as we like to reuse before we recycle, this is one case where *fresh is best* so that fungal outbreaks do not attack your collection.

Save any newspapers (without colour pictures) from your recycling.

When using printed newspaper, be sure to follow the drying times. Newspaper ink is acidic and can react with or degrade the specimen. My favourite is when a specimen is sticky (like a bulb or algae) and it's kept in the press for too long, and you can actually read the print on the specimen!

Blotter paper

Not an absolute necessity, as it is a bit pricey, but it does help to dry specimens more evenly and quickly, which results in better colour and ultimately a better specimen.

How to Press Your Plants

The crucial decision of plant pressing is about what you want the specimen to look like once dry and mounted. Where on the plant you have collected are the most important features? Will the plant be easy to mount? Will it be easy to study this specimen without having to remove plant parts to see important diagnostic features of the specimen? This is where science meets art! Your job is to make sure the important plant features can be seen while also arranging the plant in an esthetically pleasing way. While the esthetics of a final specimen are not the primary focus for pressing a specimen, I have noticed those collectors who I consider to be masters combine both the science and the art and have given us some of the most informative, long-lasting and beautiful specimens.

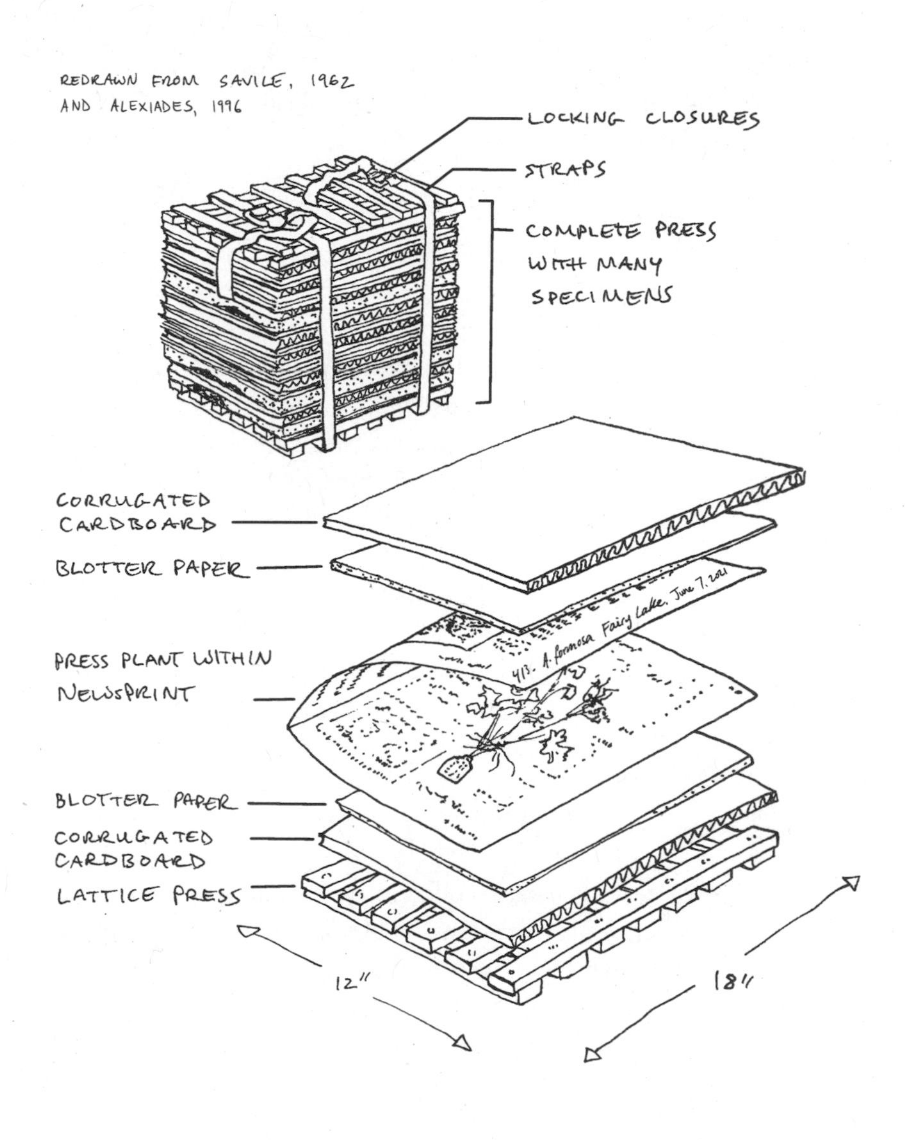

Plant press layers and a complete plant collection in a press

Here are the steps to pressing your plants:

- ✓ Take one of your two lattice press backings, and lay it down as your base layer.

- ✓ Take two to three pieces of corrugated cardboard, and lay them on top of the lattice backing to add support.

- ✓ If you want to speed up the drying process, add blotter paper to this layer to increase the extraction of moisture from the plant. I often use blotter paper during the spring when there is higher moisture and humidity, when collecting plants in wet environments (lakes, ponds and wetlands), or if I am pressing succulent or juicy plants.

- ✓ Now lay your newsprint on top of the cardboard or blotter paper, open it up lengthways and place the collected plant within it. Carefully arrange the plant (more on this shortly) with its plant tag if it has one (with the unique collection number written on it) within the newsprint, and then close it.

T If the plant is especially moist or wet from rain, dew or humidity, you can dry it off first before pressing with paper towels or include paper towels within the newsprint on top of the plant for the first few days of drying.

- ✓ Next, write the basic collection information—collection number, simple location, collection date and preliminary species identification—on the outside edge of the newsprint or on a piece of paper you include with each plant. (Make sure your collection number is unique and corresponds to your field notes.)

- Lay a second blotter paper (if using) and one piece of corrugated cardboard on top to finish the pressing for your first plant in the press. Every few specimens throughout the press, adding two to three pieces of cardboard instead of one will provide rigidity.

- Next, put consistent but light pressure with your hands or forearms across the whole pressed plant to start the process of breaking down the cells, encouraging the plant to relax within the press.

- Put the next sheet of newsprint on top of your previous piece of corrugated cardboard, lay a specimen within it, arrange, and close; record collection information; and then lay the next pressing cardboard on top, and again apply pressure.

- Repeat this process until all your specimens are pressed.

- When you are finished pressing your set of plants, place your second lattice press backing on top of the stack, and wrap the two straps around the whole press near each end.

- Pull down on the straps, resulting in a stable, level and tight press, but not too tight—hand-pressure strength but not a forceful pull as this can ruin your specimens.

- And finally, move your press to its drying location.

When arranging a plant for the pressing process, you should consider these points:

To show both sides of a large leaf means you might need to pick off a few leaves and press them separately, and then mount them on the same sheet showing features of both sides of the leaf. You can also fold or twist a leaf while it's still attached to the plant.

If you have no flowers to spare, you can write a note in your field book to transfer to the specimen label the important diagnostic features for identification, including what you see with your hand lens.

You can take these extra plant parts and press them within the same newsprint (if it fits) or arrange them on the next newsprint with the same collection number.

* Make sure all the important features of the plant can be easily seen.
* Make sure both sides of any plant part can be seen, especially the underside of leaves. The hairs are very informative and can be quite different from the top side.
* Whenever possible, show the inside of at least one flower to help with identification. If the inside features of the flower (pistil, stamens and petals, and hairs and colour patterns for pollinators) are hidden once pressed, you will not be able to confirm these important features later when identifying the plant. This does not mean you have to cut open every flower on the specimen. I suggest cutting open one to two flowers and pressing them on the same sheet. You can place them in a fragment packet when it comes time to mount the pressed plant (see "How to Mount Your Pressed Plants," page 55) so that all these important features can be seen without having to risk destroying your specimen later.
* Try not to overlap leaves and flowers. You can remove parts if they are repeated often or overlapping, making a specimen uninformative. But, and this is important, if you do remove any parts, make sure you leave a trace amount of the part still attached,

like a small bit of stem from a large leaf, ensuring you don't take off so much that you hinder future specimen identification.

* When arranging and pressing your specimens, make sure to stay within the boundaries of the mounting paper size of 29 × 42 cm (11½″ × 16½″). The size of herbarium paper is mostly standardized around the world, and therefore any part of your plant that is pressed outside these boundaries will likely have to be removed for mounting.
* If you have plants that are taller than 42 cm (16½″), and many are, there are a few different ways to press them. You can bend them in the shape of a V or W to fit on the herbarium sheet, or you can cut them into sections and mount on separate sheets (with the same collection number). In our herbarium I call this a multi-sheet to signify that the whole specimen is on multiple sheets.
* If you have plant parts that are quite thick, such as roots, bulbs or large fruits, you can cut them into halves or thirds, aiming for a thickness of between 5 and 7 mm (3⁄16″–¼″). This can help with identification and mounting and will facilitate faster drying.
* You can use commercially available foam (about 4 cm/1½″ thick) for thicker specimens, especially those with woody stems. Place it on one side of your pressed plant on top of the newsprint when closed. The foam moulds around the stem and presses the leaves flat. The downside is that the foam can inhibit drying, but if there is a good source of heat, the resulting specimens will be flatter and more useful.
* Only put one individual plant of each species in one sheet of newsprint. If your

T If you're working with sticky or juicy plant parts, a layer of wax paper (or parchment paper) between the plant feature and newsprint will discourage sticking to the newsprint.

plant species is small, then you can collect two or three (or many if tiny) from the same population (growing close together) and press within one sheet of newsprint. These will also have the same collection number.

* Remove as much dirt as possible from the roots without damaging them.
* Leave mosses and lichens. These can be added to the fragment package as they provide valuable habitat information.

How to Dry Your Pressed Plants

Exsiccate: to remove moisture from, to dry

"Why not just crank up the heat and dry it really fast?" Great question! We not only want to make a specimen that is scientifically useful and esthetically beautiful, but we also want them to last hundreds of years beyond our lifetime. If we crank up the heat, we will probably lose many of the colours we are trying to retain for identification (especially blue pigments like in *Campanula*), as well as much of the intact DNA that has become so useful in research. The trick is to have enough but not too much heat!

The key factors for successful drying and pressing are consistent air flow, temperature, air humidity and even pressure across all specimens. The temperature of the heat source, air circulation through the corrugated cardboard, thickness of the plant material collected and conditions they were collected in (very wet or humid), and what other plants are in the press will all affect the drying rate. Find the best setup in your house to optimize drying factors (air flow, temperature and humidity). Place your press near a heating source (not touching—safety first) with a fan nearby to circulate the heated air through the whole press. I have a small space heater and a rectangular large fan that work quite well for this purpose.

On average I would say most specimens take from three to six days to dry (or more if they are thicker or more succulent) at a temperature of 24°C–27°C (75°F–80°F).

The Crucial Drying Week

Now that all your plants are in the press, don't forget about them! Remember this next step of drying is where patience and care are a must. Many collectors think that once a plant is in the press, it's out of sight, out of mind, but that is not the case. You should check your pressed plants every few days to ensure they are drying at an even rate across the whole press.

- ✔ After one day in the press it's time to have a look at the plants, as they are now more flexible to arrange into place. Be careful as the specimen is often attached to the newsprint. Use tweezers to pull the plant away from the newsprint.

- ✔ Every two to three days, open the press and switch out the blotter paper or corrugated cardboard that is immediately beside the specimen if moist to the touch. Think of the newsprint, blotter paper and cardboard as a protective sponge. These papers are extracting moisture from the plant, but once the sponges are full, they need to be replaced. Otherwise, your plant will keep trying to reabsorb the water from these papers, which can cause wavy or wrinkled plant pressings, fungal infections and *dermestid* (bug) outbreaks.

"How will I know when the pressed plant is finished drying?" Feel the outside of the pressing paper. If it is moist, do not open; continue to dry. You know a specimen is dry when it becomes stiff and brittle. If you can easily pick up the plant and it keeps the pressed shape and feels stiff, then you can tell it is ready to mount. That said, a plant can easily reabsorb water from the humidity in your house, so make sure to properly store your specimens (see "Storing Your Collection," page 65).

- ✔ When you close your press again, expect to tighten the press by a few centimetres (about an inch) more than before to ensure your specimens will continue to compress.

- ✔ Repeat these steps until your specimens are dry. Remember, not all plant pressings will be dry at the same time. Be sure to remove any specimens from the press that are dry and set them aside.

- ✔ Continue drying the rest of the plants in the press until they are all finished drying. Do not stop the process of drying until it is complete and ready for the next stage—mounting.

4 Mounting Your Pressed Plants

I see the mounting phase as the true artistic stage of the process. This is where it all comes together—the pressed plant and the collection information telling the story of how this became a magnificent specimen. Mounting is also a relaxing and enjoyable experience. My favourite days in the University of British Columbia Herbarium happen around the big work table, sharing the mounting experience with volunteers as we also share stories of our day, talking life, favourite foods and classes. It's a "mounting circle." There is also something so satisfying about taking a crisp, plain white sheet of herbarium paper and securing a pressed plant, the label and stamp to create a complete and hopefully timeless specimen.

While there are alternatives that are less expensive for mounting supplies, if you want your collections to last for centuries, I encourage you to use archival, acid-free, and pH-balanced supplies. You can find the paper, glue and tape in many art supply stores or from library, archival and herbarium supply companies.

Mounting Supplies

Corrugated cardboard
Herbarium mounting paper
Archival glue in a bottle
Archival tape and scissors
Wet sponge
Paper towels
Tweezers
Herbarium labels
Fragment packets
Weights or beanbags

Corrugated cardboard
You can use the corrugated cardboard from your press or use another set specifically for mounting.

Herbarium mounting paper
Herbarium mounting paper comes in a standard size of 29 × 42 cm (11½″ × 16½″) and is made from cotton fibres only. This paper is much thicker than normal printer paper, an important characteristic since the paper needs to support a plant's weight and last hundreds of years. The paper also has a neutral pH of 7.0–8.0; if the paper were acidic, this would degrade the specimen over time.

Archival glue in a bottle
Archival glue—white polyvinyl acetate (PVA)—dries clear, is acid-free, does not emit harmful odours, is water soluble and

does not become brittle as it ages. Avoid any chemicals or toxic glues that will negatively interact with the specimen, your collection or yourself. It is important that the adhesive is water soluble, so if needed, you can add water to the specimen later to lift a plant part to study. It is also important that the adhesive does not become brittle with age, as this can lead to large leaves shattering on the paper or parts of plants breaking off. Over time, your plant and the paper will expand and contract with changes in humidity, but at different rates—so use a glue that will stay flexible.

Archival glue might seem expensive, but it is often quite thick and can be thinned with a few drops of water to your liking. Many craft glues and even white glues can be used, but they just won't last as long.

Archival tape and scissors

Also known as gummed linen, archival tape is used for mounting your pressed plant to the mounting paper. This tape is water activated and dries very quickly, which means you don't have to wait for any glue to dry. You can buy these precut, but I use scissors and cut my own pieces for each mounting.

Wet sponge

Use a wet sponge to dampen the archival tape before you put it onto the specimen and paper, as honestly, it just doesn't taste that good!

If you do not have archival tape, you can always hand-sew (with cotton thread) large parts of the specimen that are still lifting off the paper after the glue has dried.

Paper towels

Paper towels are useful for wiping off any fresh excess glue without leaving too many fibres behind.

Tweezers

Tweezers are useful to grab and move plant parts and help weave the archival tape between plant parts when mounting.

Normally we type and print our labels, but nothing is stopping you from handwriting each one. Just make sure it is legible, as this is for history!

Herbarium labels

The label includes all the information about your collected plant and is just as important as the plant itself. I encourage you to use 100% cotton paper instead of regular printer paper, which is often made from tree pulp and can be very acidic. Our label size is normally 10.8 × 9.3 cm (4.3″ × 3.7″)—an 8½″ × 11″ sheet cut into six labels—but you can pick whatever size label you want.

Fragment packets

Fragment packets are small paper envelopes useful for holding small plant material, like flowers and leaves that have been removed in the field or have fallen off the specimen and can't be reattached. Fragment packets can also be helpful for containing messy aquatic or many little tiny plants as these can be time consuming and difficult to mount. You can buy pre-made packets from herbarium supply companies, but at the herbarium we custom-make them, depending on the size we need, with the same cotton paper we use for labels.

You can also make little beanbags of different shapes and weights to lay on top of parts of your specimen. There are many tutorials online.

Weights or beanbags

These can be handy when you need to weigh the plant down to the paper while the glue dries. At the herbarium, we use little rectangular metal weights that are 2.5 × 12 cm (1″ × 4.75″) and also a few circular weights measuring 5 cm (2″) in diameter.

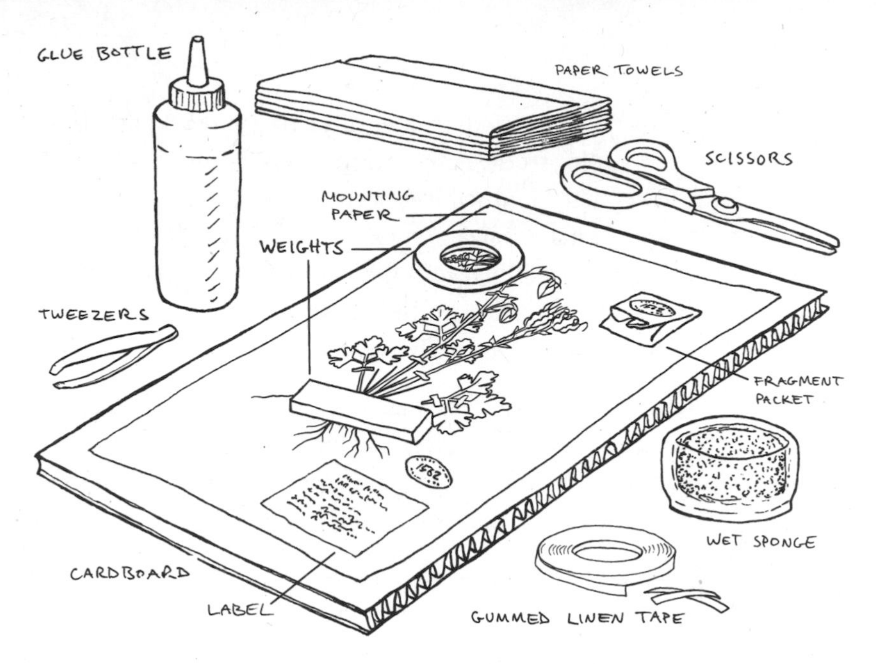

Mounted pressed plant with weights and mounting supplies

How to Mount Your Pressed Plants

Over the years I have seen specimen mounting done in a variety of ways for a variety of reasons. I tend to use a few different styles that work for the variation of specimens that come to our herbarium.

Before mounting begins, you will need to consider space for all the pieces that make up a specimen: your pressed plant and its label, the fragment packet, the collection stamp and any future annotation labels that will be added if you donate your specimen to an institution. The fragment packet is where you store any extra plant material, such as detached fruits, small flowers and extra leaves, and can be fastened on the sheet wherever they fit best. A collection stamp gives ownership to that specimen and is located above the specimen label. An annotation label describes when a researcher has examined, re-identified or removed parts from the specimen for study. Annotation labels are usually located in the lower, left-hand section of the mounting paper.

There are many ways to mount a specimen using glue, tape and sewing, all in differing combinations depending on what works best for you. Below is a description of how I mount specimens with my team of volunteers.

- ✔ First, I take a clean herbarium mounting sheet and place it on a piece of corrugated cardboard (usually one of my pressing cardboards) as a backboard for each mounting to ensure the plant is well supported if I have to move it.
- ✔ Next, I place my pressed plant so all the key characteristics can be seen, including both sides of a leaf and the inside of one flower (if possible), leaving space for roots and accessory fruits.
- ✔ To keep my plant away from the edge, I tuck in all the leaves, flowers and plant parts and leave a 2 cm (¾″) boundary around the whole sheet. Remember, these specimens will be handled over and over again, so leaving an edge will help ensure specimen longevity.
- ✔ Any parts that are hanging off the sides (this often happens with grasses) will eventually break off, so I remove them now and put them into a fragment packet.
- ✔ When I arrange my plant, I make sure the important reproductive features are the most protected. I try to avoid having flowers and fruits and important plant features (for identification) near the edge of the paper. This might mean I arrange the plant upside-down on the paper. Researchers don't care what orientation the plant is mounted, as long as the specimen is useful and protected.

- You will also learn how to "balance" a specimen on a sheet, with the heavier portion near the lower half of the sheet, as this prevents the sheet from bending while being handled. However, in a case where many thick specimens (e.g., an alpine cushion plant) of the same species are being mounted, I mount them in different positions on the paper. Otherwise, the folder they are filed into becomes unbalanced and sloped, which is not good for long-term storage.
- Once I have my plant or plants arranged on the mounting paper, it is time to secure them. I use a combination of glue and archival tape. If the plant has a bulky feature like large root bases, thick stems or large, thick leaves that will not be easily held down by a few strips of tape, I place glue in the middle of the plant part and turn it over to place the plant on the paper.
- I prepare little pieces of archival tape into strips of varying size, depending on what I am mounting. My tape sizes are usually around 2–4 mm × 2–3 cm (⅛″ × 1″).
- I moisten the gummy (shiny side) of the strapping tape and place the strip tightly over the stem, securing it to the paper on both sides, and hold down on the paper for 30 seconds, locking the stem to the paper. Be careful not to break your specimen when you are strapping it down!

Do not tape across more than one stem or plant part, except with particular groups like grasses. One big tape strap across a bundle of stems will never hold the specimen in place, and the plant will eventually slip out of the straps and slide off the paper.

- I usually start at the bottom of the plant's main stem and work my way up, taping the plant to the paper. The number of tape straps you use on a specimen and how you strap them is very dependent on the bulk or thickness of the plant and how the plant is arranged once dried.

- Next, I sparingly cover my label with glue and place it in the bottom right-hand corner, leaving about 2–3 mm (1⁄16″–1⁄8″) from the edge of the mounting paper. This label placement is consistent throughout collections, as it makes it easier to flip through specimens when they are stored inside their protective folders.

- I then mount the fragment packet in the upper right-hand corner of the specimen sheet, 3 cm (1¼″) from the edge.

- When the specimen is lying flat on the mounting table, if any plant parts have shadows (meaning they are raised a little higher above the paper), I put a small dab of glue here and there to secure these parts.

- You can add one to two weights to these sections if the parts keep popping up or won't lie flat. Keep them there until the glue dries.

Note: Be careful that there isn't any glue sticking to the weight, as you wouldn't want to tear off some of the specimen or paper when you remove the weight.

- At the end of the mounting, once the glue is dry, I lift the herbarium sheet with the plant vertically and do a gentle shake test to see if any plant parts are lifting off the paper. If so, I again add either a strap or a dab of glue.

The key to mounting is to make sure you are securely strapping the plant to the paper so it can be safely stored while at the same time not taping over any important features needed for identification. This means not taping over the leaf tips, buds or flower heads. Also, do not apply any cross strappings, as strappings should not overlap.

Any future annotations will be glued at the bottom of a specimen sheet, so leave about 5 cm (2") room at the bottom of the rest of the mounting paper.

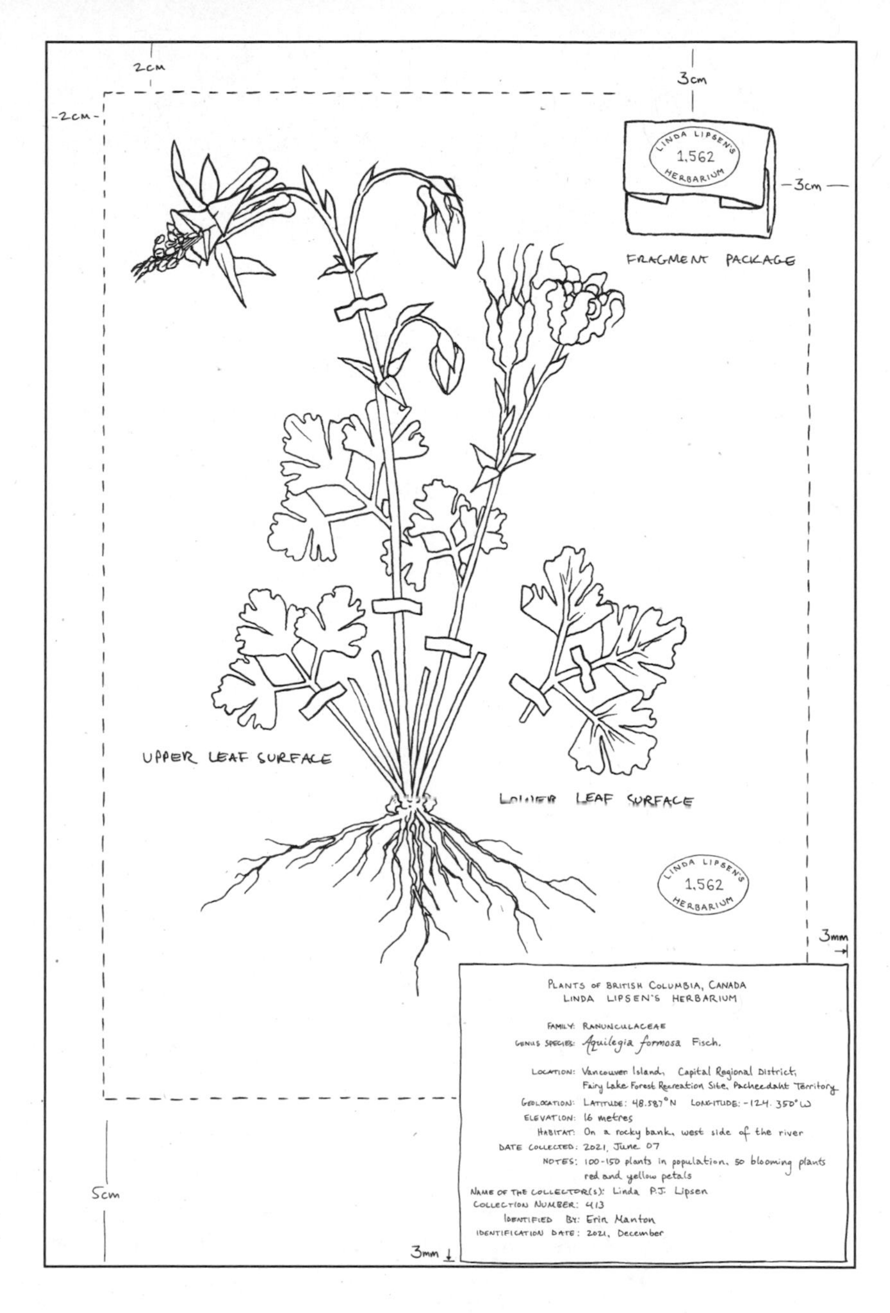

Finished specimen

✔ Lastly, I stamp our herbarium stamp just above the specimen label to show ownership of this specimen.

DIFFICULT PLANTS TO MOUNT

Large plants

There are times when the specimen will be so large, it fills the mounting paper. In this case just glue the right-hand edge of the label so it is secure on the mounting paper but the viewer is able to see the whole specimen by flipping the label out of the way. If the plant is larger than one mounting sheet, you can mount on two sheets, which is known as a multi-sheet. Sometimes collectors mount the reproductive parts on one sheet and the vegetative parts on the second. Take care to not remove any important identification features if you are removing parts to make the specimen fit.

Thick plant parts: Roots, fruits and thick stems

If you have a thick part, I would encourage a large dab of glue on the "back" and then one or two horizontal straps on either side of the glued part to hold it in place.

Lots of tiny plants

I do love little plants, but mounting multiple tiny individuals can take quite a bit of patience. Use your tweezers to separate each individual, then carefully place small dabs of glue on the backside of each plant. Add one or two small straps to fully secure the plant, usually one on the root and the other on the stem of the flower(s). If the tiny plants are a mishmash of plant parts, like is often found in aquatic plants, or are just so tiny that you can't strap them, I suggest using a larger packet (similar to a fragment packet). The specimens will be loose but secure inside the closed packet.

Unwieldy grasses, sedges and rushes

I have been mounting for over half my life, and I can tell you now, no one likes mounting grasses. They are unwieldy and all over the place and just won't stay put, and they have a way of slipping out of mounting tape, like Houdini! It takes some real patience to glue and strap down all the stems and then try to make them look esthetically pleasing. Due to their height, they are often bent and pressed in a V shape so the whole plant can fit on a sheet. I suggest you put a big wallop of glue on the thick root base. Though my guidance earlier was to not clump stems together, with grasses I do often clump three to five stems and strap them down on the mounting paper together, until all the stems have been mounted in clumped sets—otherwise, it will take all day to do one mounting. For grasses, do not apply glue or tape at the point where the leaf joins the stem. This area contains the ligule, which is an important taxonomic character that must be visible.

This is a great opportunity to have a creative moment: mount the plants to form a shape like a heart or circle or even spell something out on the sheet.

Overlapping flowers or leaves

Sometimes a plant has an abundance of flowers and leaves, but they are so layered that you can't strap an individual plant part. In this case, you have a couple of options. You can remove a few of the plant parts and put them into a fragment packet or use a second sheet to mount them separately to show clearly a full flower or leaf shape. If you do the latter, make sure to leave a bit of the plant part you remove on the original specimen so it is obvious a plant feature was there.

Conifers, trees, shrubs and cones

Conifers can be quite thick and stiff, and their cones can be large, even very large (some weigh up to 3 kg/7 pounds)! For needled conifers, like pines, I use strapping tape to adhere them to the mounting paper. For scaly conifer leaves, like cedars, I use glue on the leaves and adhere them to the paper. If the cones are not too bulky or heavy for the paper, you can mount them with the specimen. But some cones will need to be removed from the collected plant and put into a storage box. You should make a separate label and tag for the cone so it will be clear which cone is with which sheet-mounted specimen. There are some conifer species, like hemlocks and spruce, where all the needles will fall off unless chemically treated. In this case you gather up all the needles and put them into a sealed plastic bag and then into a large fragment packet. Many cones will also have parts that fall off, as they are meant to disperse the plant's seeds, so you can treat them in the same way: in a plastic bag inside a fragment packet (or inside a storage container).

Ferns

When identifying ferns, it is crucial for identification to see both sides of a leaf (frond) including its reproductive features, which are known as *sori* (round dots or continual lines) and are usually found on the underside of the frond. Either bend the leaf halfway along, or mount two leaves, one facing up and the other facing down. I use some strapping tape up the middle of the frond around the mid-vein, and then use glue sparingly throughout the rest of the frond.

5 Preserving and Organizing Your Collection

Now that you have put all this hard work into your collection, how will you take care of it? To preserve your collection for the long term, you must freeze and store your specimens properly. I have seen thousands upon thousands of specimens thrown away because of insect damage, light degradation, mould, or dust issues. If you follow just these few simple steps, your collection can last hundreds of years. I often tell people we have no idea how long a cared-for collection will last or how useful yours will become, so take good care of your specimens.

Freezing Your Collection

By freezing your collection, you will deter insects from eating your beautiful work. What many forget is that in collected plants, insects have often previously laid their eggs in the open flowers. These may hitch a ride into your collection too! When their insect larvae hatch, they will easily find pollen and seeds to feed upon. The two most damaging insects are carpet beetles (*Attagenus unicolor*) and cigar beetles (*Lasioderma serricorne*). The larvae of both species can eat a whole flower head, even through the mounting paper, leaving a hole in your sheet where there used to be a lovely flower. The other two insects that often find their way into plant collections are silverfish (*Lepisma saccharina*) and book lice (*Psocoptera* sp.). Neither of these species will attack your plant per se, but both will feast on the paper and glue, easily gobbling up a label and all its important data. Silverfish are often found in damp, dark places (like basements), while book lice burrow into almost any container you might be storing your collection in.

We freeze our unmounted specimens before we bring them into the herbarium, thus preventing any critters from entering into the area where plants are being mounted or studied. We will freeze them again once they are mounted and before we put them into the cabinets for long-term storage.

To freeze your dried, mounted specimens:

- ✔ Place your mounted specimens within a large folded piece of newsprint or thick paper, then put a second piece of folded paper on the opposite side to enclose the specimens securely on both sides.

- ✓ Sandwich the specimens between two pieces of pressing cardboard, and tie string around the cardboard edges (similar to your press straps) to secure a tight bundle.

- ✓ Place this bundle into a large plastic bag, and securely seal the bag by tying a knot and then taping it closed. Leave this bag of specimens in your home freezer (–18°C/0°F) for 10–14 days depending on the number of specimens in the bundle and the size of the plant parts.

- ✓ When you remove the specimens from the freezer, leave them sealed inside the bag until there is no moisture or condensation on the outside of the bag. The amount of condensation is fairly small after the plants have been properly dried and should dissipate within 24 hours. You can now remove these specimens from the bag and organize them in their storage container.

- ✓ Repeat this freezing activity for your collection every one to two years to deter any new insect infestations. Once an outbreak starts, they can eat a whole collection and ruin your life's work!

Storing Your Collection

You should store your collection where it will be protected and accessible. Try to find a place that is dry, with relatively consistent, low air moisture (humidity), a cooler temperature (18°C/65°C) and preferably in the dark, as the less exposure to UV light, the better. If your storage area has high humidity, your plants or paper could reabsorb moisture from the air, leading to moulding, insect outbreaks and warping.

Ideally, store your beautifully organized specimens in some type of cabinet or box system for long-term storage. This could be in a closet or shelving unit (with doors). Just make sure your specimens are protected from being damaged by bumping or getting squashed by other items placed on top. If you do use boxes, try to use archival, non-acidic boxes, as regular cardboard boxes will disintegrate the edges of your collection sheets over time.

Organizing Your Collection

You can organize your collection based on whatever system you like: blooming times, habitat type, year collected, plant uses, even flower colour. It's your collection, and you can do as you please! In the herbarium world, we tend to organize the collections based on evolutionary relationships—that is, by family, genus and species. This makes it relatively easy to find a specimen, which is essential when you have millions of sheets to track.

Geography is another layer you can eventually add into your organizational system. Since we use specimens to map species distribution around the globe, when your collection is large enough, it often makes sense to add this layer. Once we put our specimens in our database, the data is shared with a variety of web portals such as the Global Biodiversity Information Facility (GBIF), where anyone can search Earth's biodiversity data. Organizing by geography will support a better understanding of where species are found and where they have gone missing, facilitating more comprehensive biodiversity research.

Once you decide how you want to organize your collection, you should place the same set (e.g., species) of specimens inside a lightweight, non-acidic (archival) species folder. We take all the species folders and place them together into a genus folder made of thicker card stock. The species and genus folders are both labelled at the bottom of the folders. In our herbarium we distinguish geography with a colour-band on the genus folders. The specimens are fully covered by both folders (with no edges sticking out), as over time UV light will damage plant colour and disintegrate paper edges if they are exposed.

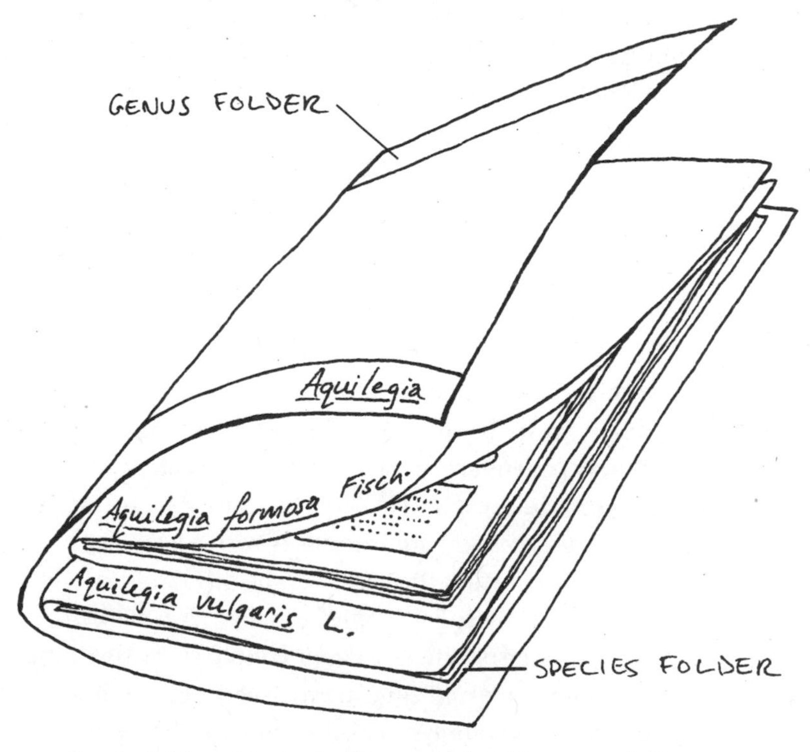

Species folders alphabetically arranged within a genus folder

Cataloguing Your Collection

While your specimen labels and collection notebook are your primary source of information, more often than not, once you start organizing you will want to move your collection information into some sort of electronic database, like a spreadsheet. This will allow you to cross-reference your whole collection, which can be quite powerful. You can then easily reference all the locations where you collected, what years you collected, what species you collected, how many times and in what types of habitats. A digital system also allows you to add information that might not be with the specimen itself, such as accessory fruits you collected, notations about DNA samples collected, your field pictures or a note about having mounted a specimen on multiple sheets. A personal collection database will also add a second layer of protection to your collection data.

Accessioning Your Collection

The last stage for any collection is *accessioning* your specimen. In my herbarium we all like accessioning as it means the specimens are ready to join our collection. At the University of British Columbia, since we have five distinct collections, the accession number starts with a letter to indicate which collection the specimen belongs to (V = vascular, B = bryophyte, F = fungi, L = lichen, A = algae). You may accession specimens to your collection by stamping your herbarium name and entering the specimen's unique accession collection number. This specimen will now have a unique collection number given in the field and a unique accession number given when it was incorporated into your personal collection. Accessioning your personal collection is optional, but if a collection is large enough and the collector is very active, accessioning can make a lot of sense. For instance, it can be an advantage if you incorporate other people's specimens into your collection. Or if a specimen becomes damaged and must be removed, you will still have a record of its existence in your accession records.

Donating Your Collection

Many collectors ask me if they can donate their specimens to our herbarium, and they are surprised when I say YES! This is an easy way to preserve your collection, especially beyond your lifetime—simply donate it to a provincial or university herbarium or similar facility. The benefits of doing this would be proper care (freezing every year or two) and correct climate control, but most importantly, your specimen would be accessioned and available for worldwide research and teaching. If you would like your collection to survive well beyond your lifetime, donating to an institution with the expertise to manage specimen collections is the best approach. These donated specimens can be valuable to the herbarium and to science.

6 Identifying Your Specimen

Let me start off by saying that there are an estimated 300,000 flowering plant species in the world. This number does not include conifers or ferns, which make up another 15,000 species. As a comparison, there are estimated to be 30,000 *tetrapods* (four-legged vertebrates) and 5,400 mammal species on Earth. Putting a name to a plant is no small task. It will require time and patience, but it is well worth the effort.

It's all Latin and Greek to me! Latin and Greek were originally chosen for scientific names as they are stable and reliable and are not easily influenced by other languages. Useful for reading but difficult for speaking. People can be sticklers for pronunciation, but don't fret. There is no one right way to pronounce Latin words; after all, it's a dead language, so no one can be too sure how exactly it should be said. As long as *we* (the people discussing this plant) each know what species we are discussing, then we have both done our job to communicate clearly, even if we may be mispronouncing it.

To better communicate and catalogue this grand organismal diversity, the Swedish botanist Carl Linnaeus, in the 18th century, developed a naming system known as *binomial nomenclature*. In this system each species is usually identified with two words, the genus and the specific epithet or species. A taxonomist will name and describe a new species according to the International Code of Botanical Nomenclature to hopefully provide a meaningful and stable name. While life on Earth and its classification is much more complicated than this, we still use the base idea of this naming system designed by Linnaeus.

How to Identify Your Specimens

Most people are worried about this stage of collecting, the accountability stage. Do you know what you collected? Can you identify your specimens? Of course you can! This is just like any other hobby or skill. To get good at it you have to practise, and luckily that's pretty easy as you can look at flowers and plants almost everywhere as you go about your day. When you are waiting at a bus stop, going for a walk or run, gardening or at the park, *stop and smell the roses*—and then count the flower's parts (number of sepals, petals, stamens and pistils)! Have a look for some fruit, and check out the leaf arrangement, shapes and anything else notable. There is more to it than that, but that is where identification begins.

Identification Resources

Field guides are a great place to begin, as many experts put their knowledge into field guides to encourage the learning of local flora. These books often focus on species you would commonly find within your region's flora, and many have extra information about closely related species that you might confuse your particular species with.

The next stage would be to use a larger and comprehensive regional flora. A *flora* is a taxonomic guide to every known plant found within a region, which often includes taxonomic illustrations of key plant features for identifying a plant. The keys and specialized names for plant parts might take some time to get used to as it does start to feel like you are learning a whole new language (and you pretty much are). But with practice, you will better understand the diversity of plants around you and the key features that distinguish them from one another.

Many websites and applications are fast evolving and there are a lot to choose from, and by the time this book is printed there will be another 100 more! What I have relied on for years is regional online portals of biological data like EFlora BC (ibis.geog.ubc.ca/biodiversity/eflora/) or the Consortium of Pacific Northwest Herbaria (pnwherbaria.org). These sites support specimen data, specimen images, species distribution maps and curated photos with identifications you can trust. iNaturalist is another useful tool that generates an ID for photos you submit, and experts can confirm your identification. You too can help become a resource by contributing photos, locations and specimens and make these resources even better through community science.

Key Features of Large Families and Distinct Plant Groups

Below are a few notes about the important parts of plants belonging to some of the larger plant families. I've also included some thoughts about a few interesting groups that you are sure to run across in your collecting adventures. Once you've identified your specimen to a family, then you can consider the more detailed classifications of genus and species.

Key Features of Large Plant Families

Below, the common family name is given first and then the scientific family name in brackets ending in *aceae*. Some families have had their name changed over the past years, and this former name is also included (what you might find in older reference books). Plant terms in *italic* are often specific to that family or group and important to note and collect.

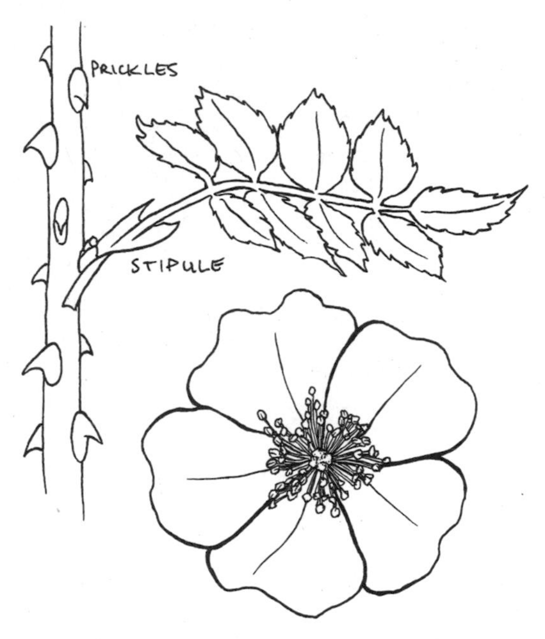

Rose (Rosaceae)

This diverse family ranges from trees to shrubs to herbaceous plants. Their flowers always have more than 10 stamens (also true of the buttercup family) and typically five petals. When collecting specimens of this family, bring your gloves and be careful of all the *prickles, thorns* and *spines*. Unfortunately, these painful parts are needed for identification. Make sure you show their *stipules*, usually two little leaflets at the base of a leaf stem. Of course, their flowers are important to collect, and their fruits are often needed to confirm identification.

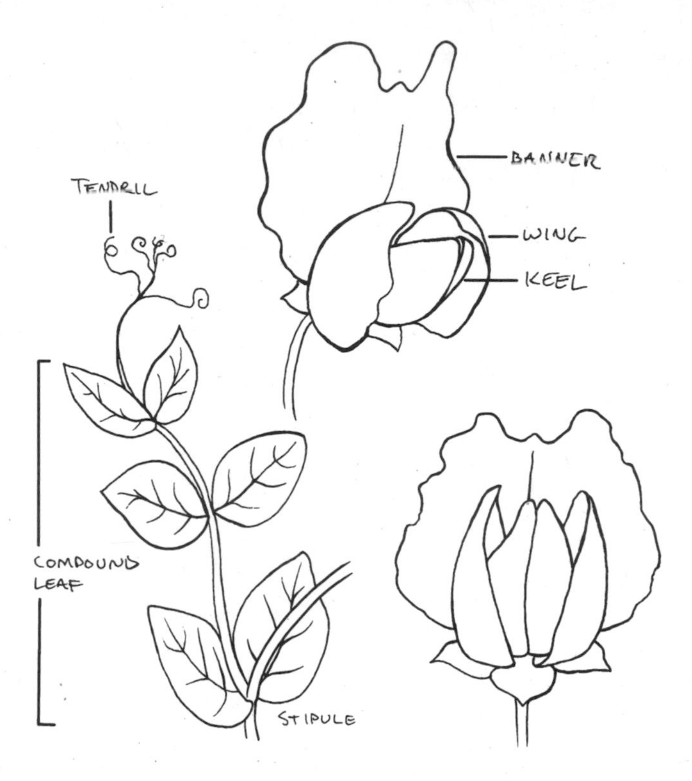

Pea (Fabaceae, formerly known as Leguminosae)
The pea family is known for their complex flowers of five petals that can be described as one *banner*, two *wings* and two *keels*. The keels are fused together along the bottom to form a boat and house the reproductive structures under pressure as a pollinator mechanism. It is important to collect the *compound leaves*, and don't forget the *tendrils* (very cool modified leaves for grabbing and climbing). *Stipules* also occur in this family. Make sure to collect both flowers and mature fruit pods. Sometimes the fruit pods will be ripening lower on the stem while new flowers emerge at the tippy-top part of the same stem.

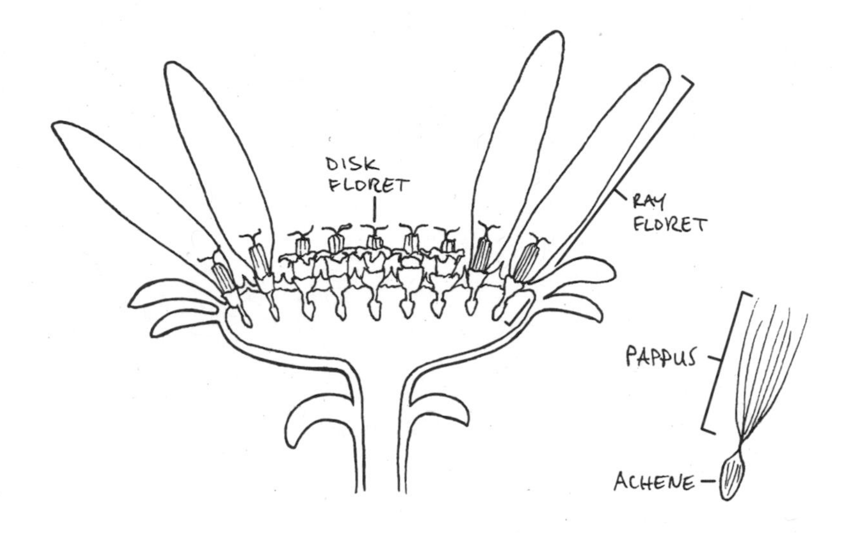

Sunflower (Asteraceae, formerly known as Compositae)
The flower heads of the sunflower family are actually a collection of flowers, or *florets*. Plants within this family are often classified by their two different floret types: the tubular-shaped inner *disk florets* and the strap-shaped outer *ray* or *ligulate florets* that are typically thought of as the petals. Some species have both types of florets, and some have only one. Their seed-like fruits are known as *achenes*, and the white fluffy parts of dandelions we love so much for making wishes are really for dispersing seed by wind and are known as the *pappus*.

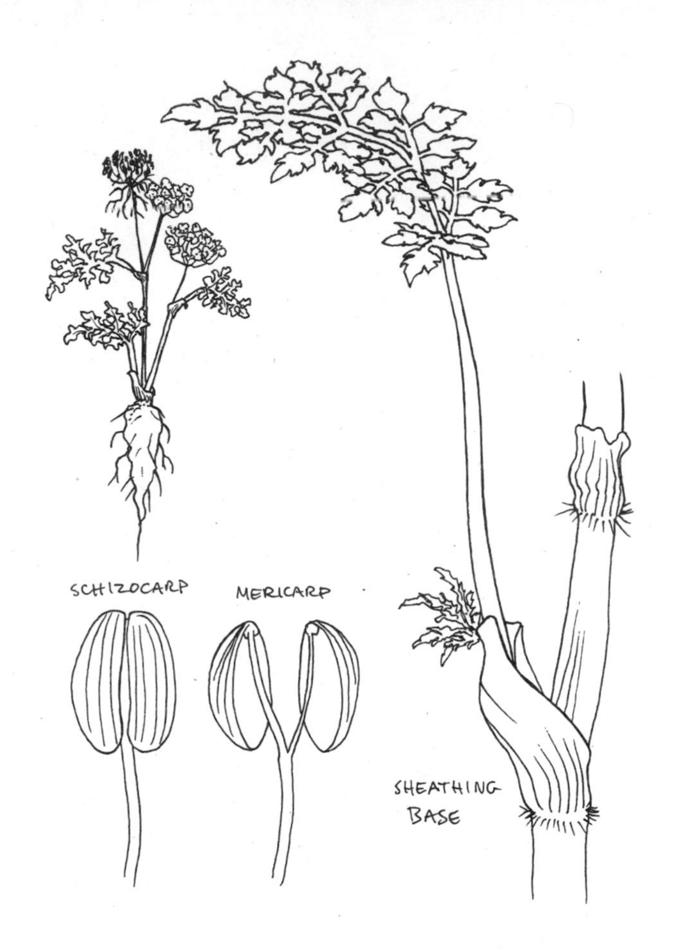

Carrot (Apiaceae, formerly known as Umbelliferae)
This family has umbrella-shaped inflorescences (a collection of flowers). Stems are important, especially showing the compound leaves with their *sheathing bases* that wrap around the main stem. This is a family where *roots* will be needed for identification, as some are pointy like a carrot and others are *fibrous* or *tuber*-shaped like a beet. You will also need flowers and mature fruits known as *schizocarps*, which split into two *mericarps*.

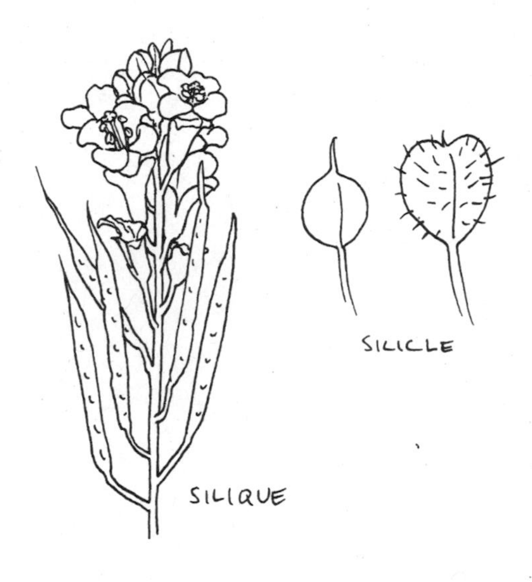

Mustard (Brassicaceae, formerly known as Cruciferae)
This family has four-petalled flowers that look like an X. (Species of the evening primrose family also have four petals.) The mature fruits are especially important to collect for this family as plant identification keys often start off with the shape of the mature seed pods. If seed pods are long, they are known as *siliques*, and if they are short, they are known as *silicles*. Leaf type, shape and their diversity of hair types will also be important when identifying. Remember to collect basal leaves for this family.

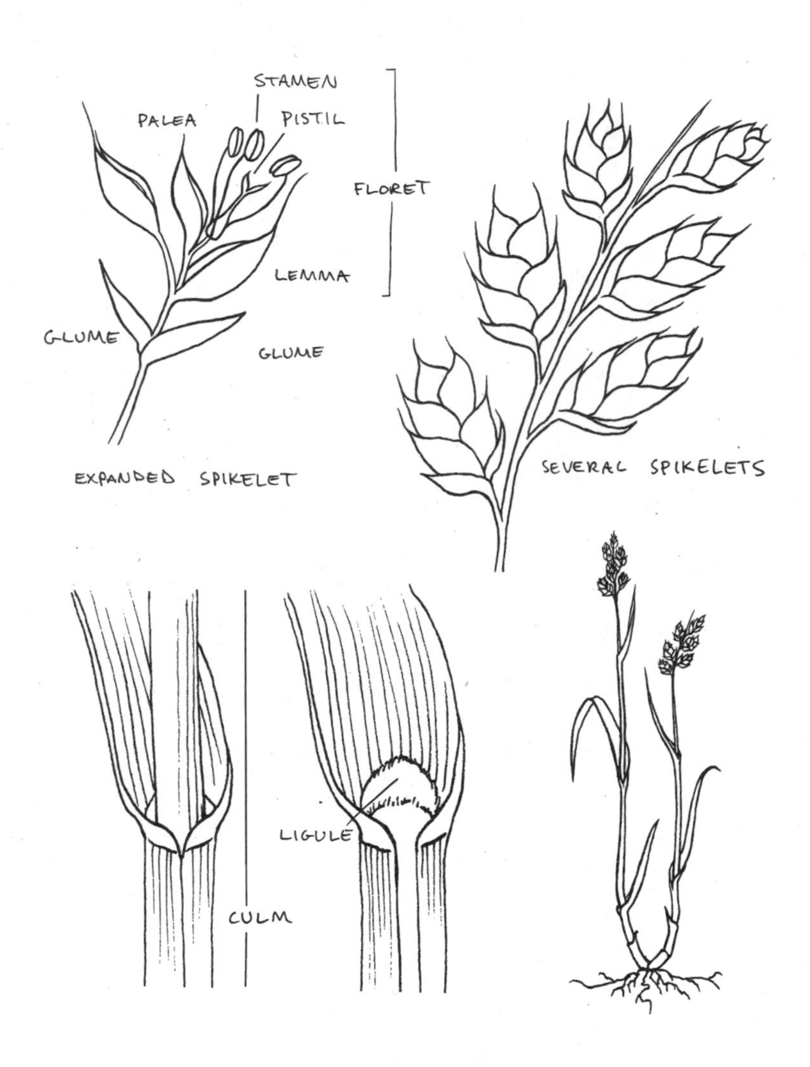
STAMEN
PALEA
PISTIL
FLORET
LEMMA
GLUME
GLUME
EXPANDED SPIKELET
SEVERAL SPIKELETS
LIGULE
CULM

Grass (Poaceae, formerly known as Graminaceae)
This is one of the most economically important plant families in the world, including all the grains. Make sure the plants are in bloom. Grasses do not have showy flowers. Specialized terminology describes the structures that surround the reproductive parts (the pistils and the stamens). A single pistil is enclosed within a *lemma* and a *palea* and is called a *floret*. One or more florets are enclosed within (usually) two *glumes* and is referred to as a *spikelet*. A hand lens will be helpful. Remember, it's important to collect the *culm* (the grass stem—especially the flowering ones) and leaves with their distinctive *ligules*. Often the stamens only last a few days, so if you do see anthers—bonus. For this group you better bring out the hori hori as you will want to dig up their roots.

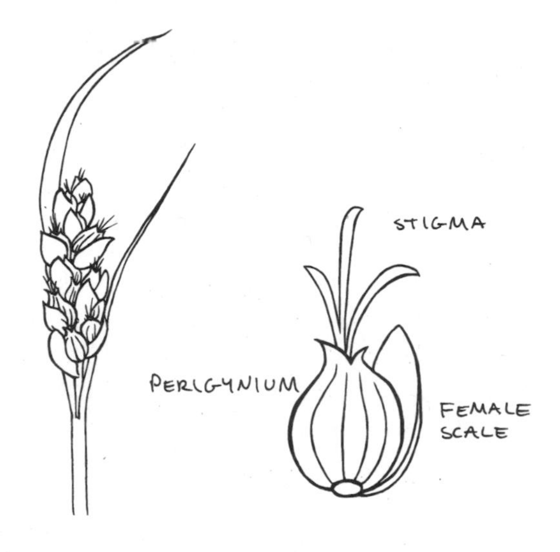

Sedge (Cyperaceae) and Rushes (Juncaceae)

The saying goes "Sedges have edges and rushes are round and grasses are hollow to the ground" (except at the nodes). Both of these groups look like grasses but have distinctly different floral structures from grasses. When you collect either of these groups, you need to make sure to collect them in flower. Sedges must be collected in the mature *fruiting* stage (after the flowering stage) to show the *perigynium*, or membranous sac that surrounds their seed, along with the size, shape and colour of the *female scale* (surrounding leaflike structure). You will also want to dig some of the root system to show if they have rhizomes.

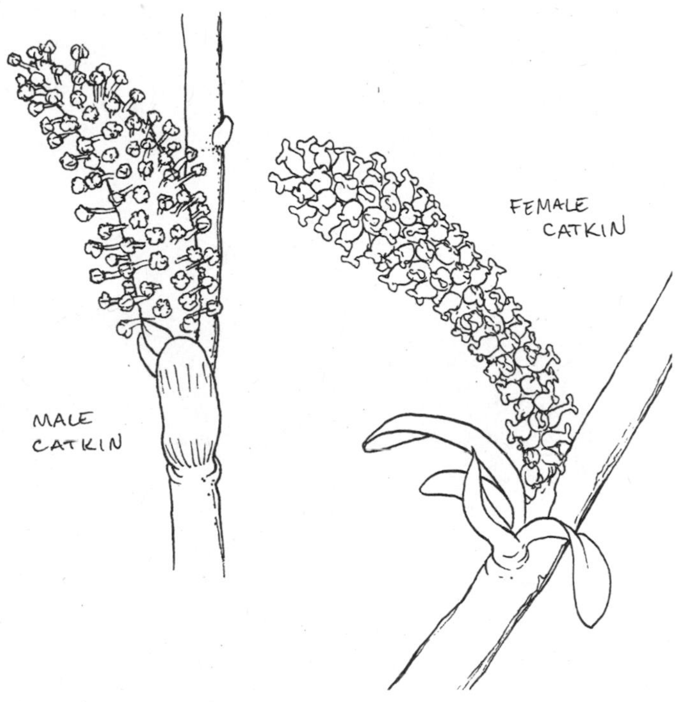

Willow (Salicaceae)

Oh, I just love collecting willows—but identifying, not so much! Willows remind me of spring, beautiful waterways and fluffy soft flowers. But this group is tough to identify, and you must ensure you collect a mature flower *catkin*, preferably both a male and a female (which occur on separate plants), as well as a leafy stem.

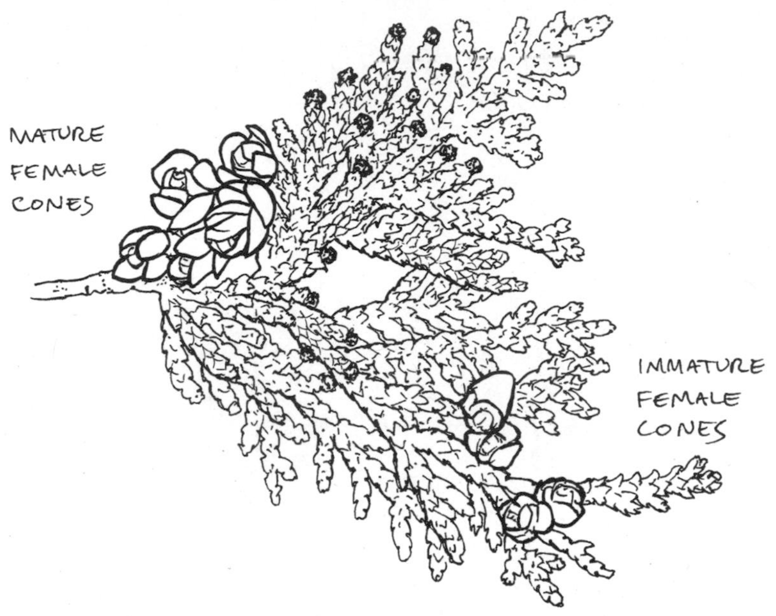

Scale-like conifer with male and female cones

Needle-like conifer with mature female cone

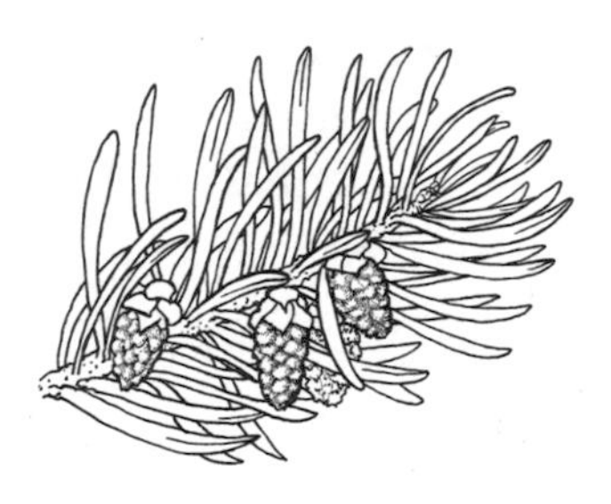

Needle-like conifer with mature male cone

Key Features of Distinct Plant Groups

Conifers

Collect a typically shaped branch with their needles or scale-like leaves, and try to get both the conspicuous mature female cones and more inconspicuous male cones (quite small and usually found at the tips of branches). If you are lucky with timing, you can collect this year's maturing cones along with last year's mature cones, which also shows how much that branch grew in the year!

When I first started collecting, I thought, how am I supposed to get up this tree for a sample? A wise collector told me that most trees, including conifers, drop their branches from age or a windstorm, and you can often collect suitable branches lying under the tree. Branches from higher up in the tree show their morphology in a compact form (perfect for a sample) as compared with the wider-spreading understorey branches.

Ferns

Collect mature ferns. Immature ferns can be difficult to identify, as their important leaf traits often do not show themselves until complete maturity. Make sure to collect when the ferns are in the reproductive stage where you see lots of little dots or continuous brown lines on the underside of the leaf. These are known as *sori* (reproductive parts).

Fern underside with "little dots" sori

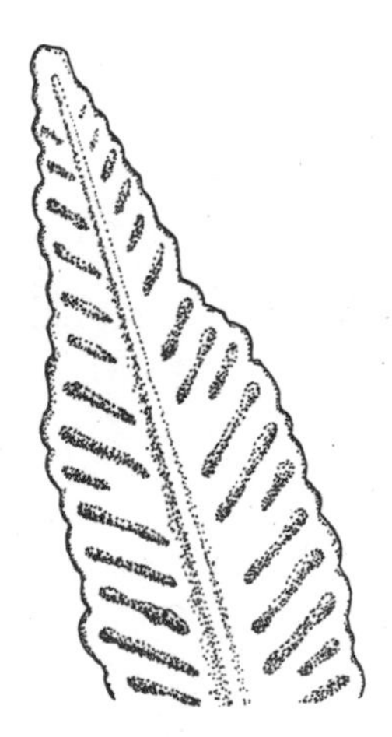

Fern underside with continuous sori lines

References and Resources

Plant Collecting and Preservation

Brayshaw, T. Christopher. *Plant Collecting for the Amateur*. Victoria, BC: Royal British Columbia Museum, 1996.

Bridson, Diane, and Leonard Forman. *The Herbarium Handbook*. 3rd ed. London, UK: Royal Botanic Gardens, Kew, 1998.

British Columbia Ministry of Forests. *Techniques and Procedures for Collecting, Preserving, Processing, and Storing Botanical Specimens*. Work. Pap. 18/199. Victoria, BC: Province of British Columbia, Ministry of Forests Research Program, 1996.

Clark, Lewis J. *Wild Flowers of the Pacific Northwest: From Alaska to Northern California*. Edited by John G. Trelawny. 2nd ed. Sidney, BC: Gray's Publishing, 1976.

Herbarium Supply Company, herbariumsupply.com

Hicks, Arthur James, and Pearl M. Hicks. "A Selected Bibliography of Plant Collection and Herbarium Curation." *Taxon* 27, no. 1: 63–99. doi.org/10.2307/1220483.

Liesner, R., comp. *Field Techniques Used by Missouri Botanical Garden*. St. Louis, MO: Missouri Botanical Garden, 1995. mobot.org/MOBOT/molib/fieldtechbook/welcome.shtml.

"Preparation of Plant Specimens for Deposit as Herbarium Vouchers" (web page). Accessed July 12, 2022. University of Florida Herbarium, Florida Museum. floridamuseum.ufl.edu/herbarium/methods/vouchers/.

Raven, Peter H., and Tamra Engelhorn. "A Plea for the Collection of Common Plants." *New Zealand Journal of Botany* 9, no 1: 217–22. doi.org/10.1080/0028825X.1971.10430177.

Savile, D.B.O. *Collection and Care of Botanical Specimens*. Publication 1113. Ottawa, ON: Canada Department of Agriculture, 1962.

Thiers, Barbara M. *Herbarium: The Quest to Preserve and Classify the World's Plants*. Portland, OR: Timber Press, 2020.

Specialized Groups of Plants

Ceska, Adolf, and Oldriska Ceska. "More on the Techniques for Collecting Aquatic and Marsh Plants." *Annals of the Missouri Botanical Garden* 73, no. 4: 825–27.

Haynes, Robert R. "Techniques for Collecting Aquatic and Marsh Plants." *Annals of the Missouri Botanical Garden* 71, no. 1: 229–31. doi.org/10.2307/2399065.

Ethical Collecting

"BC Species & Ecosystems Explorer." Conservation Data Centre, Province of British Columbia. Accessed April 30, 2022. www2.gov.bc.ca/gov/content/environment/plants-animals-ecosystems/conservation-data-centre/explore-cdc-data/species-and-ecosystems-explorer.

Birchweaver, Adam. "The Honourable Harvest: Guiding Principles to Restoring Our Relationship to the Natural World." The Gaia Project, April 12, 2021. thegaiaproject.ca/en/the-honourable-harvest-guiding-principles-to-restoring-our-relationship-to-the-natural-world/.

Wall Kimmerer, Robin. *Braiding Sweetgrass: Indigenous Wisdom, Scientific Knowledge and the Teachings of Plants*. Minneapolis, MN: Milkweed Editions, 2013.

Wall Kimmerer, Robin. "The 'Honorable Harvest': Lessons from an Indigenous Tradition of Giving Thanks." *YES! Magazine*, November 26, 2015. yesmagazine.org/issue/good-health/2015/11/26/the-honorable-harvest-lessons-from-an-indigenous-tradition-of-giving-thanks.

Plant Identification

Douglas, George W., Del Meidinger and Jennifer L. Penny. *Rare Native Vascular Plants of British Columbia*. Victoria, BC: Conservation Data Centre, 2002.

Douglas, George W., Gerald B. Straley, Del Meidinger and Jim Pojar, eds. *Illustrated Flora of British Columbia*. 8 vols. Victoria, BC: BC Ministry of Environment, Land and Parks; BC Ministry of Sustainable Resource Management; and BC Ministry of Forests, 1998–2002.

Flora of North America Editorial Committee. *Flora of North America*. 28 vols. New York: Oxford University Press, 1993–2014.

Hitchcock, C. Leo, and Arthur Cronquist. *Flora of the Pacific Northwest: An Illustrated Manual*. Edited by David E. Giblin, Ben S. Legler, Peter F. Zika and Richard G. Olmstead. 2nd ed. Seattle, WA: University of Washington Press, 2018.

Kozloff, Eugene N. *Plants of Western Oregon, Washington & British Columbia*. Portland, OR: Timber Press, 2005.

Pojar, Jim and Andy MacKinnon, *Plants of the Pacific Northwest Coast*, Rev. ed. Tukwila, WA: Lone Pine, 2016.

Turner, Mark, and Phyllis Gustafson. *Wildflowers of the Pacific Northwest*. Portland, OR: Timber Press, 2006.

Websites

Beaty Biodiversity Museum herbarium. beatymuseum.ubc.ca/research-2/collections/herbarium.

Consortium of Pacific Northwest Herbaria. pnwherbaria.org.

E-Flora BC: Electronic Atlas of the Flora of British Columbia. ibis.geog.ubc.ca/biodiversity/eflora.

Flora of North America. floranorthamerica.org.

Royal British Columbia Museum herbarium.royalbcmuseum.bc.ca/collections/natural-history/botany.

Index

Illustrations are indicated by page numbers in bold